Experimenting for Sustainable Transport

Transport, Development and Sustainability

Series editor: David Banister, Professor of Transport Planning, University College London

European Transport Policy and Sustainable Mobility

David Banister, Dominic Stead, Peter Steen, Jonas Åkerman, Karl Dreborg, Peter Nijkamp and Ruggero Schleicher-Tappeser

The Politics of Mobility: transport, the environment and public policy

Geoff Vigar

Experimenting for Sustainable Transport: the approach of Strategic Niche Management

Remco Hoogma, René Kemp, Johan Schot and Bernard Truffer

Transport Planning Second Edition

David Banister

Remco Hoogma
René Kemp
Johan Schot
Bernhard Truffer

Experimenting for Sustainable Transport

The approach of Strategic Niche Management

Routledge
Taylor & Francis Group
LONDON AND NEW YORK

First published 2002 by Routledge

Published 2017 by Routledge
2 Park Square, Milton Park, Abingdon, Oxon OX14 4RN
711 Third Avenue, New York, NY 10017, USA

Routledge is an imprint of the Taylor & Francis Group, an informa business

Typeset in Sabon and Imago by NP Design & Print, Wallingford, Oxfordshire

British Library Cataloguing in Publication Data
A catalogue record for this book is available from the British Library

Library of Congress Cataloging in Publication Data
A catalog record for this book has been requested

ISBN 978-0-415-27117-2 (pbk)

Contents

Remco Hoogma is a Programme Manager at NOVEM, responsible for the electric and hybrid vehicle programme. NOVEM is the Dutch organization responsible for the implementation of government programmes in the field of energy and the environment. At the time of the research for the book he was employed as a research fellow at the University of Twente's Centre for Studies of Science, Technology and Society. He has a broad research expertise in transportation, environmental and technology issues, and has recently finished a PhD thesis on the subject of technological niches for electric vehicles.

René Kemp is a Senior Research Fellow at the Maastricht Economic Research Institute on Innovation and Technology (MERIT), Maastricht University and Research Director of Science, Technology and Economic Policy (STEP) in Oslo. He is an expert on environmental policy and technology, a topic on which he has published extensively and on which he is consulted by the European Commission and OECD. His book *Environmental Policy and Technical Change* is an authoritative work on the subject. His current work focuses on system innovation and transitions to sustainability. He has published in economic, policy science and social science journals. His work on transition management is adopted by Dutch environmental policy-makers.

Johan Schot is Professor in Social History of Technology at Eindhoven University of Technology and University of Twente. He is Scientific Director of the Foundation for the History of Technology and Programme Leader of the National Research Programme on the *History of Technology in the Netherlands in the 20th Century*. He is chairing (with Ruth Oldenziel) the European Science Foundation Network *Tensions of Europe, Technology in the Making of Twentieth Century Europe*. His research work and publications range from history of technology, science and technology studies, innovation and diffusion theory to constructive technology assessment, environmental management and policy studies.

Bernhard Truffer is a senior scientist at the Swiss Federal Institute for Environmental Science and Technology. He has published on innovation-oriented environmental policy in general and on transport in particular. He studied user experiences with electric vehicles and organized car sharing in Switzerland. Currently he is heading a large Swiss research project on the definition of Green Power standards in liberalized electricity markets.

This book grew out of a dissatisfaction with the way insights gained from technology studies, evolutionary economics, constructivist sociology and history of technology are put to use in the policy realm, or in fact, are hardly used. It is our belief that scholars must transform their insights into enlightening and flexible tools that can be taken up by practitioners working in governments, in businesses, NGOs and other organizations. The philosophical justification for most technology policy activities has remained unchanged for decades: because various forms of market failure lead to underinvestment in research and development (R&D), governments must stimulate such investments. An additional role often assigned to government is to educate users and wider publics to accept and embrace new solutions coming from R&D investments.

In our view, over-emphasis on issues of market failure and user acceptance has resulted in the neglect of other new and promising technology policy options, especially in the area of sustainable development. Here, new theoretical insights underscore the importance of shaping technological content, enabling the articulation and construction of new user needs and defining and negotiating the course towards sustainable development. These processes are neither purely technical, nor simple exercises in the diffusion of knowledge. Effective technology policy, in our view, is an open learning process, a series of experiments with the introduction of new technologies. Our book develops a tool – Strategic Niche Management – to help think about these learning processes. It offers a number of suggestions on how to set-up experiments. It also gives the reader access to theoretical insights, and case studies of actual experiments showing how these insights help to understand and explain what is going on.

This book has a long history. The idea of Strategic Niche Management developed out of a research programme developed at the University of Twente in the Netherlands collaborating with MERIT, University of Maastricht in the mid-1990s. Strategic Niche Management is part of an attempt to understand better technical change and its relationship to economic and societal changes. This goal is simple: to help various actors in society to build more constructive relationships with new technologies, saving them from the naive belief in either the transformative myth of the 'technical fix' or the destructive criticism that new technologies cannot be part of any solution.

This book grew out of a project supported by the European Commission, DG XII, within the area 'Human Dimensions of Environmental Change' of the

RTD Programme 'Environment and Climate', and we wish to express our gratitude for the support. The starting point of this project was the observed under-utilization of available transport technologies that might contribute to sustainable development. We focused on sixteen case studies of actual experiments. The project involved a team of excellent researchers. We would like to thank Kanehira Maruo, Birgitta Schwarz, Gerhard Prätorius, Karl-Heinz Lehrach, Sylvia Harms, Stephen Potter, Benoît Simon, Ben Lane and Matthias Weber for their ideas, and high quality research. Outside the group directly involved in the research, we also received valuable ideas and encouragement from Alex Grablowitz, Matthieu Craye, Per Sørup, Boelie Elzen, and in the final phase David Banister and two reviewers helped us through.

The project was not only developed through research but also through discussions with potential users, actors involved in the experiments we studied. We would like to thank explicitly the members of the practitioners' group for their often stimulating discussion, we refer to Joris Benninga, Daniela Cocchi, Martin Kleemeyer, Stefan Liljemark, Torbjörn Waldenby, Thomas Waschke, Urs Muntwyler and Jonathan Parker, our liaison in Brussels. The SNM approach was further tested at the conference 'Strategic Management for Clean Innovative Transport Technology and Sustainable Mobility' held in Seville, Spain, 8–10 June 1998, in collaboration with transport operators, manufacturers, policy-makers and planners of demonstration projects. We would like to thank them all for their valuable input. This conference resulted in a small book to help practitioners implement the idea of Strategic Niche Management, entitled *Experimenting with Sustainable Transport Innovations. A Workbook for Strategic Niche Management* written by Matthias Weber, Remco Hoogma, Ben Lane and Johan Schot.

After the research was done, and the discussions were over, it was time to go back to the study and write the book we intended to write. At the start we believed it would take us half a year. Now we know, it took us two long years and many meetings in between. In 2000 we thought we would never make it to the end, our lives continued with many other projects and obligations and it was difficult to pursue our path. Here the help of David Kirsch turned out to be crucial. He offered to help us write and edit the book creating more continuity between the various chapters. David clarified our thinking in many ways and also sustained us with his faith that the book would make a substantial contribution to the field of technology studies and to the practice of managing new technologies in society.

When everything is done and said, and the book is written, we are ready to give it to reader. We hope that our enjoyment of working on the idea and practice of Strategic Niche Management comes through!

Remco Hoogma
René Kemp
Johan Schot
Bernhard Truffer
November, 2001

ASTI	Accessible Sustainable Transport Integration, project in London
ATG	AutoTeiletGenossenschaft, a Swiss car-sharing organization
BART	Bay Area Rapid Transit
BEV	battery-powered electric vehicle
BFE	Bundesamt für Energie, the Swiss federal office for energy
CARB	California Air Resources Board
CAT	Clean Air Transport, an EV-developer from the early 1990s
CNG	compressed natural gas
CCT	Camden Community Transport
CGEA	Compagnie Générale des Eaux (now Vivendi)
CGFTE	Compagnie Générale Française des Transports et d'Exploitation, a French transport company (now part of the Vivendi group)
CT	community transport
DC	direct current
DAUG	Deutsche Automobilgesellschaft, the Rügen electric vehicles project leader
EC	European Commission
ECS	European Carsharing, association of several national car-sharing organizations in Europe
EDF	Electricité de France
ENTRANCE	ENergy savings in TRANsport through innovation in the Cities of Europe, a European research project
EU	European Union
EV	electric vehicle
GM	General Motors

GPS	Global positioning system
INRETS	Institut National de Recherche sur les Transports et leur Securité, the French national transport research institute
INRIA	Institut National de Recherche de l'Informatique et d'Automatisme, the French national information technology institute
LEV	lightweight electric vehicle
LPG	liquefied petroleum gas
MIRA	Motor Industry Research Association (UK)
NGV	natural gas vehicle
OECD	Organisation for Economic Co-operation and Development
OPEC	Organisation of Oil Producing and Exporting Countries
PIVCO	Personal Independent Vehicle Company (now absorbed by Ford Motor Company)
RATP	Regie Autonome des Transports Parisien, the Paris transit authority
PSA	Peugeot-Citroën S.A., French car producer
R&D	Research and development
SBB	Schweizerische Bundesbahn, the Swiss federal railways
SMH	Société suisse de microélectronique et d'horlogerie, a Swiss watch manufacturer
SNM	Strategic Niche Management
SULEV	super ultra low emission vehicle (Californian classification for cars)
TULIP	Transport Urbain Libre Individuel et Public, an individual public transport system using electric vehicles
VOC	volatile organic compounds
VCS	Verkehrsclub der Schweiz, Swiss traffic users club (mainly for public transport and bicycle riders)
VDEW	Vereinigung Deutschen Elektrizitätswerke, the association of German electric utilities
VSE	Verband der Schweizerischen Elektrizitätsunternehmen, the association of Swiss electric utilities
ZEV	zero-emission vehicle

Chapter 1

Technological fixes

When we visit an 'autosalon' or 'motor show', we are invited to enter the future of transport: progress envisioned in terms of technology and technological advance. In their celebration of technology as key to a better future, autosalons are the utopian self-images of our technological culture. In such prophesying about the future, the belief in technology seems endless. Indeed the continuing stream of new automobiles and features presented at autosalons gives technological progress a sort of inevitability. The leading idea is that whatever the transport problem, technology can fix it. Autosalons capitalize on a pervasive ideology in our culture, one called by the historian Howard Segal 'technological utopianism'.[1] Technological advance is seen, by definition, as something positive and desirable. It is based on faith in technology not merely as tools and machines, but also as the means of achieving a 'perfect' society in the near future.

The technological utopians of the early twentieth century believed that technological advances held a solution to many of the world's major problems: hunger, disease, scarcity and war. Although nobody would be so blindly optimistic today, the idea of a technical fix still holds. For example, Amory Lovins and Ernst von Weizsacker in their book *Factor 4* collected 50 examples of technological options for quadrupling resource productivity while doubling wealth.[2] Building on the idea of an eco-efficiency revolution, a Dutch government minister stated that 80% of all environmental problems could be solved through use and development of new technologies.[3] We see the technical fix paradigm as one of the articles of faith of modernity and of our modern technological culture. It continues to dominate many policy debates on the future of transport and of our society.

While transport technologies seem to fulfil dreams, the twentieth century has shown that technology also creates nightmares. Each year 250,000 people are killed in road accidents and 10 million people are injured. Fossil fuel based transport is also a slow killer through the release of toxic compounds (e.g. CO, lead, benzene and other VOCs) and the release of fine particulate matter that endangers human health and impairs ecosystems. Congestion limits mobility and creates economic costs through lost time. There is a long list of other problems: noise, large land use, energy dependence, and the undermining of communities.[4] And the negative impacts of transport are likely to worsen. The business-as-usual scenario of the OECD Environmentally Sustainable Transport study projects that by 2030 car ownership and total distances travelled will increase by up to 200% in many OECD countries, with even greater increases predicted in the freight and aviation sectors. Although technical improvements have led to significant reductions in emissions per (tonne) km, these trends will lead to increased emissions levels for most transport pollutants in 2030. The highest increase in automobile emissions will be for CO_2, which is likely to double by 2030, given foreseeable policy action.[5] The overall conclusion of the OECD is that likely advances in technology will not be sufficient to overcome increased emissions; transport in 2030 will be moving *away from* rather than towards environmental sustainability.

At the World's Fair of 1939 in New York, General Motors' Futurama pavilion presented quite another transport future. Nearly 28,000 people visited each day, waiting for hours to enter the pavilion. When it was finally their turn, visitors were led to a row of 'sound chairs' fixed to a moving conveyor. A fifteen-minute journey to the future then began, as seen from a plane flying low over a scene of America in 1960. The very first words heard by each traveller were: 'Strange, Fantastic? Unbelievable? Remember this is the world of 1960'. In their journey, visitors saw the transport world of the near future – cruise-controlled superhighways, frictionless, clean and free from accidents.[6] As we now know, the future turned out differently.

The technical fix paradigm is not, however, the only ideology present in our own culture. Forceful as it has been, a voice critical of technology has always been present as well. Technology has been referred to as the 'Megamachine' (Lewis Mumford) and 'Autonomous Technology' (Jacques Ellul). Jane Jacobs has drawn attention to the negative effects of the automobile on American cities.[7] These critics portrayed technology as part of the problem and argued that it could not be seen as part of the solution. They stressed the need for value changes that would lead to drastic reductions in demand for mobility. This opposing view can be seen as a call for a cultural or social fix instead of a technical fix. Despite many proclamations since the late 1960s about the impending death of the automobile, the automotive paradigm is still very much alive.

This brief review introduces two very different approaches, both of which are deeply embedded in our culture and dominate the debate about the future of transport. The first, the technical fix ideology starts with the strong conviction that advances in transport technology will always – in the end – yield benefits that vastly outweigh the costs and that technology itself should never be the subject of policy debates. The only reason for policy-makers to become involved in technology policy is because of market failures leading to under-investment in research and development (R&D). In this view, R&D

subsidies are the appropriate answer to contribute to the development of new technologies that solve transport problems.

Proponents of the second, the cultural fix paradigm, view technology as part of the problem. The only way out is *not* to start with technology. Real solutions will have to come from social and cultural change. Restricting mobility (through price mechanisms or by providing mobility quotas) is the way forward. At the same time, technical change needs to be regulated through standard setting. In the final analysis, for those following the cultural fix ideology, the central question seems to be whether the human family has the moral and political will to develop new and stringent constraints on mobility patterns. If the answer is no, then we better prepare for Autogeddon.

The key issue addressed in this book is whether we are indeed trapped within these two contending positions. What seems very difficult to do is to raise social and cultural issues in relation to technology. We tend to call for either a technical or a social fix.[8] Here we will develop an approach, termed Strategic Niche Management, which would allow for working on both the technical *and* the social side in a simultaneous and coherent manner. The perspective is that new transport technologies could contribute to solving transport problems if they are introduced in a socially embedded way. We argue that this approach is especially suited and needed in order to develop a more sustainable transport future since such a future will inevitably include both social changes (e.g. of mobility patterns, of perceptions of mobility etc.) and technological changes (e.g. the introduction of new means of mobility).

Recent technology studies have shown that the social and technical are always mixed up in designs, artefacts and technical systems. Technology has a hybrid character.[9] First of all, designers have social visions and values that influence – often implicitly – their technological choices and designs. Second, when technologies are introduced in society, consumers, the general public and policy-makers all influence the integration of those technologies into daily life and in this way determine the final fate of any transport technology. Consumers are not passive users of innovation; on the contrary, they shape user patterns often in unexpected ways.[10] The car culture was not made solely by Ford and General Motors, but also by American consumers. This is one of the reasons why the GM panorama at the World's Fair of 1939 did not in fact come to be! Consumers did not accept automated control and slow driving.

This feature of technological change, what could be called the co-production (or co-evolution or co-construction) of the technical and the social, has not been recognized well enough in the debate about the future of transport. Hardly any policy instruments try to exploit and work upon the socio-technical features of transportation systems. When co-production is recognized, it immediately becomes clear that a major issue is how to create technical and social solutions simultaneously. How can we create synergy between the social and the technical? How can we embed new mobility values and patterns in material realities and vice versa? How can we create new technologies which invite the emergence of new transport modes? In this book we propose to develop the new policy perspective of Strategic Niche Management that tries to do just this.

The approach of Strategic Niche Management

Central to the Strategic Niche Management (SNM) policy perspective is the view that technology policy must contribute to the creation and development of niches for promising new technologies through experimentation. The support should be temporary and not too strong so as not to create 'Mama's boys'. A niche can be defined as a discrete application domain (habitat) where actors are prepared to work with specific functionalities, accept such teething problems as higher costs, and are willing to invest in improvements of new technology and the development of new markets. If successful, a technology might move to follow-up niches resulting in a process of niche branching. Eventually, that technology might compete head-on with the dominant technological option in a part of its market or markets.

The proposition that it is possible and productive to engage in Strategic Niche Management (SNM) rests on two fundamental assumptions. The first assumption is that the introduction of new technologies is a social process that is neither an unavoidable deterministic result of an internal scientific and technological logic, nor a simple outcome of the operation of the market mechanisms. This assumption has been explained above with the notion of co-evolution or co-production. The second assumption is that it makes sense to experiment with this co-evolutionary nature of technology. Such experiments can be envisaged as (part of) a niche in which technologies are specified and consumers are defined and concretized. Experiments make it possible to establish an open-ended search and learning process, and also to work towards societal embedding and adoption of new technology.

In the area of transport there is no lack of experiments, top-down ones organized in demonstration projects, or bottom-up ones developed in small market niches. They are abundantly present, and in this book we will discuss eight examples in depth: experiments with electric vehicles; on-demand public transport services (dial-a-ride); car sharing; bicycle pools using electronic access reservation and billing systems; and self-service public vehicles. Demonstration projects are not only present in the transport area. For example, in the field of energy technology, organic farming, and in the multi-media area (where users experiments with new forms of information and communication technology), we also see actors exploring new options through experiments. Therefore, although we concentrate on the transport example, our conclusions are of broader interest. We propose a new kind of technology policy for sustainable development.[11]

To stress the idea that learning is central in SNM we prefer to use the notion of experiment rather than notions such as demonstration or pilot projects. This learning goes beyond technical learning; it involves learning about user needs, societal benefits and negative effects, and regulation – and not just learning to specify existing user needs, technological options and regulatory requirements (i.e. forecasting), but also learning to question existing preferences and to find ways of building new ones. SNM is not just about testing user acceptance but tries to find ways for tinkering with user needs. SNM, thus, stresses what we have called above the co-production of the technical and the social. In the transport case, this implies development of new technologies connected to new mobility forms and regulatory incentives. User needs, specific technological options, regulatory requirements and demands must be developed in relation to each other. Experiments can provide platforms for such development processes.

From our analysis of eight experiments, it will become clear, however, that they are often not geared towards exploring and exploiting how new technological opportunities can offer new ways of providing mobility and sustainability. Learning about possible new sustainable transport pathways is not the central issue. Instead, demonstration of a fancy new technology seems to be more central. Moreover, demonstration projects often remain isolated events. When the experiment is finished, there is no follow-up activity. The new technologies do not seem to create a larger market or niche. Something always seems to go wrong with these experiments. The SNM perspective provides a new way of examining these efforts to identify ways of upgrading existing practices, by viewing the experiments in the context of niche development processes.

SNM and sustainable development

Technological change plays an important role in any policy directed at improving the environment. Technological innovation can change the energy and material basis of economic and societal processes and result in drastic improvements of resource productivity.[12] At the same time, it is clear that most technological change consists of incremental improvements and does not go beyond the control of particular pollutants. Ecological restructuring of production and consumption patterns will require not so much a substitution of old technologies by new ones, but radical shifts in technological systems or technological regimes (this notion will be introduced in Chapter 2) including a change in consumption patterns (user preferences), regulations, and artefacts. It is here that the SNM approach makes a contribution. Existing policies of providing economic incentives and of regulation can tackle a number of environmental problems. However, these policies are not particularly suited to deal with technological regime shifts. Typically, they have resulted in so-called end-of-pipe technologies.[13] Sustainable development has been a banner for many causes. The question can be asked to what extent technologies can be called sustainable. It is clear that this can only be the case if a broad notion of technology is used – one which encompasses user practices and regulatory institutions.

However, starting with technological change as a point of entry for discussing options for sustainable transport results in a bias, and other choices are possible. It is also conceivable, for example, to start with the emergence of new green values, new participation modes visible in local 'Agenda 21' processes, or new management practices within business. But, such starting points have their own bias and often lead to a neglect of incorporating technological changes. In this book, we have chosen to use technological change as the starting point since it is an obligatory point of passage for any movement towards sustainable development. Technical change is a central feature of modern societies and a powerful source for social change. Machines do change the world. Directing the construction of new technologies towards a more desirable path is an urgent task, especially where society lacks perspectives, instruments, and policies to accomplish this mission.

In its simplest form, sustainable development is defined as development that meets the needs of the present generation without compromising the ability of future generations to meet their own needs.[14] It assumes that it is possible, indeed necessary, to make trade-offs between continuous economic growth and the sustainability of the environment. Through the development of new modes of economic growth, pollution and resource use can be reduced dramatically. Furthermore, it is possible to deal with issues of equity and democracy on a world scale. Finally sustainable development requires that society, business and governments operate on a different time scale. Long-term aims must not be sacrificed for short-term gains. SNM makes contributions to all these elements. First, through contributing to the development of specific promising technologies, it helps to get the economy to a new growth path, which in turn might spread to developing countries that are eager to enjoy the fruits of economic growth. Technological transformation will be a primary strategy for reaching such a path. Second, SNM aims to create platforms for early user interaction with new technologies. This is a much-neglected dimension in the sustainability debate.[15] Third, SNM is about inducing long-term changes whose effects might only become visible (in terms of reduction of environmental impacts) over decades.

Can SNM thus be called a technological fix after all, albeit a socio-technical fix? It is clear that SNM involves a dilemma. On the one hand, it is driven by the concern to introduce new transport technologies and practices that would alleviate existing transport problems. In this sense, SNM does prescribe technological solutions to current problems. Diffusion is not possible without making choices, fixing certain technological and demand (mobility) options and subsequently creating new path-dependencies. So creating fixes is part of SNM. On the other hand, learning is central when applying SNM, which implies the deconstruction (unfixing) of present day mobility patterns and connected technologies and the articulation of new kinds of patterns and technologies. In order to allow for learning flexibility must be preserved to limit the possibility of premature selection of inappropriate solutions. Therefore this tension between learning and institutional embedding is inevitable and perhaps desirable. It is precisely the learning processes encouraged in an SNM approach that lead to early identification of negative impacts and explorations of technological, regulatory and user needs. They allow one to work towards solutions to problems of unsustainability beyond technical fixes.

The SNM approach puts learning processes at the forefront, with the result that it becomes difficult to be specific about outcomes beforehand. What kind of sustainable transport future will then be pursued through SNM? In a way applying SNM demands that we avoid trying to answer this question directly. We do not know in detail what a sustainable transport regime looks like, but we have well-founded insights into possible building blocks of such a regime, which include not only technologies but also new regulations, consumption patterns etc. To make a step towards sustainability these elements need to be investigated – not separately but linked. The SNM approach leads to a detour, to a process in which a sustainable transport future is explored. However, implementing the SNM approach does include picking a set of technologies for experiments, so an assessment *ex-ante* of the potential of new technologies is necessary. This assessment is not, however, focused on gains in terms of resource productivity or social equality of individual technologies. Rather, SNM tries to answer the question: which technologies might form a pathway towards a more sustainable technological regime including a new set of consumption patterns (user preferences), regulations, and artefacts?

We can also put it another way. SNM concerns *changing change*: introduction processes might be designed differently. The long-term goal of SNM policies is to create new routines ('institutions' as neo-institutional economists would call them) that would anticipate impacts, user requirements and related technical choices earlier and more frequently, to set up introduction processes to stimulate learning and reflexivity, and thus to create space for experimentation. In the long run the ability to deal with difficult and complex processes such as the introduction of more sustainable technologies and mobility concepts will become more widespread. Having said this, we do not claim that SNM guarantees sustainable development. However, uncertainty is an intrinsic part of both technical change and sustainable development and cannot be lifted through moral entrepreneurship for a certain course either.

Content of the book

The flow in this book is simple and straightforward. Chapter 2 starts out asking the question why many promising technological options – exemplified by the concept vehicles exhibited at autosalons – are not introduced into the market. Subsequently, a whole set of technological, economic, and social barriers are discussed which cannot be dealt with in isolation. What we have is a structure of interrelated factors that feed back upon each other, the combined influence of which gives rise to inertia and leads to specific patterns in technological change, especially the improvement of already dominant technological options. To account for this phenomenon, the concept of technological regime is introduced and related to other concepts present in the innovation literature. The concept of regime helps to explain why most change is not radical, that is, does not lead to major changes on both the supply and demand side. The question is raised, what is involved in technological regime shifts (or transformations)? This leads to a discussion of the dynamics of such shifts for which a process of niche development turns out to be crucial. Patterns of niche development are suggested, including network building (featuring the importance of outsiders), learning (about user needs), and specifying expectations (to get more robust support). Then the idea of SNM is introduced and developed. SNM is positioned as a new policy approach aimed at stimulating niche development patterns.

A main mechanism for this is to set up experiments. Our argument is not that experiments would make transport more sustainable in the short run, but in the long run they could make a crucial contribution if designed in a specific way. They make promising technologies more viable, help to explore options and ways to develop them through network building, and they create platforms for involving users and other parties so that they start to reconsider their mobility behaviour. Thus SNM is a process for all parties to consider what sustainable development could imply. It is an open learning process, not a top-down message.

To pick the right kind of experiment, an analysis of the transport regime is necessary. Chapter 3 starts out analysing the way we have chosen to organize our mobility function. It argues that two separate regimes have emerged for land-based passenger transport: a private, car-based regime, and a public transport regime. Within both regimes, a number of niches are identified that contain promising beginnings for the emergence of a new regime. We discuss alternative fuels, electric vehicles, transport telematics, car-sharing, integration of public transport means, self-service rental schemes, dial-a-ride services, and bicycle pools. Some of these developments are in competition, while others sustain each other. We argue that this diversity of promising niches adds up to two possible scenarios: one of regime optimization and one of regime-shifts. We identify two developments that are most likely to contribute to a regime shift because they *could* implicate a change in technology base, user patterns, and regulations: battery-powered vehicles and the development of technologies that bridge the gap between the private and public transportation regimes. We analyse why both routes, electrifying mobility and reconfiguring mobility, are promising.

Chapters 4 and 5 discuss both routes in depth. These chapters are built up in a similar way. First we put electrifying mobility (Chapter 4) and reconfiguring mobility (Chapter 5) in a historical perspective. We provide background to a new interest in both options for upgrading and transforming transport regimes. We continue with a detailed description of four experiments. In chapter 4 we discuss: (1) the Rügen project in Germany with 60 conversion design electric vehicles; (2) La Rochelle in France involving 50 conversion design electric vehicles; (3) City Bee/TH!NK purpose-built electric vehicles in Norway introduced in Oslo and California; and (4) the Mendrisio experiment aiming to introduce 350 electric vehicles in Switzerland. These experiments cover four different paths and strategies (combining two dimensions: experiments with conversion design versus purpose design vehicles, and experiments which take users' preferences for granted versus those aiming at the exploration of users needs) tried out by various kinds of actors in the 1990s.

In Chapter 5 we discuss the following experiments: (1) bicycle pooling assisted with smardcards and electronically controlled racks in Portsmouth, UK; (2) an experiment in London which aims at making public transport more flexible by providing door-to-door service; it involved the introduction of a small fleet of electric and compressed natural gas fuelled minibuses accessible to people with reduced mobility, and scheduling assistance software and vehicle tracking technologies; (3) car-sharing in Switzerland; (4) Praxitèle, a French experiment in Saint-Quentin-en-Yvelines with the aim to demonstrate the usefulness and economic feasibility of an individual public transport system based on a centrally managed car fleet. Figure 1.1 shows the locations of the experiments discussed in this book.

For all these cases we look at ways the experiments contribute to niche formation and identify important lessons for developing the idea of Strategic Niche Management. The lessons are brought together and analysed in more depth in Chapter 6. This final chapter will articulate the idea of SNM further and compare it with other policy tools.

Figure 1.1 Locations of the experiments presented in chapters 4 and 5

Notes

1 Segal, Howard P. (1987) The technological utopians, in Corn, Joseph J. (ed.) *Imagining Tomorrow. History, Technology and the American Future*. Cambridge, Mass: MIT Press, pp. 119–136. For an overview of concept cars presented at autosalons see, for example, Nieuwenhuis, Paul and Wells, Peter (1997) *The Death of Motoring. Car Making and Automobility in the 21st Century*. Chicester: Wiley, chapter 6.

2 Lovins, Amory and Weizsacker, Ernst von (1998) *Factor 4. Doubling Wealth – Halving Resource Use*. London: Earthscan.

3 Cited in Achterhuis, Hans (1998) *De erfenis van de Utopie*. Amsterdam: Ambo, pp. 365–366.

4 See Hodge, David C. (1995) Intelligent Transportation Systems, Land Use, and Sustainable Transportation. Paper presented at the ITS America Alternative Futures Symposium on Transportation, Technology and Society, Washington, DC.

5 OECD (1993) *Transport Growth in Question*. Paris: OECD. Whitelegg, J. (1993) *Transport for a Sustainable Future: The Case for Europe*. London: Belhaven.

6 Kilstedt, Folke T. (1987) Utopia realized: the World's Fairs of the 1930s, in Corn, Joseph J. (ed.) *Imagining Tomorrow*. Cambridge, Mass: MIT Press, pp. 97–136.

7 Three other examples are Canzler, W. and Knie, A. (1994) *Das Ende des Automobiles, Fakten und Trends zum Umbau der Autogeschellschaft*. Heidelberg: Verlag C.F. Müller; Kunstler, James Howard (1993) *The Geography of Nowhere*. New York: Simon and Schuster; and Kay, Jane Holtz (1997) *Asphalt Nation: How the Automobile Took over America*. Berkeley, CA: University of California Press.

8 For a fuller historical analysis of this divide in ways Western societies manage technology, see Rip, A., Misa, Th. J. and Schot, J. (1995) *Managing Technology in Society. The Approach of*

Constructive Technology Assessment. London: Pinter Publishers, especially the introduction. See also Schot, J. (1998) Constructive technology assessment comes of age, in Jamison, Andrew (ed.) *Technology Meets the Public*. Aalborg: Aalborg University Press, pp. 207–232. The subject of the great divide between the technological and the social was very convincingly presented in the work of Bruno Latour. See among others Latour, Bruno (1987) *Science in Action. How to Follow Scientists and Engineers through Society*. Cambridge, Mass: Harvard University Press.

9 See work of Latour, *op cit.*, p. 8, and MacKenzie, Donald and Wajcman, Judy (eds.) (1999) *The Social Shaping of Technology*, 2nd ed. Buckingham: Open University Press.

10 A classic case is the early telephone diffusion. Promotors at first only saw productive usages in the business sector, while consumers persisted in using it for social purposes. See Fischer, Claude S. (1992) *America Calling. A Social History of the Telephone to 1940*. Berkeley, CA: University of California Press, p. 85. For this argument see also Lie, M. and Sørensen, K. (eds.) (1996) *Making Technology our Own? Domesticating Technology into Everyday Life*. Oslo: Scandinavian Press.

11 SNM is part of a broader set of possible policies; this whole set is called constructive technology assessment. A first casebook is Rip, Misa and Schot, *op cit*. See also Schot, J. and Rip, A. (1997) The past and the future of constructive technology assessment. *Technological Forecasting and Social Change*, **54**(2-3), pp. 251–268.

12 Weaver, P. *et al.* (1999) *Sustainable Technology Development*. Sheffield: Greenleaf Publishing.

13 For an analysis of the results of environmental policy instruments, see Kemp, René (1997) *Environmental Policy and Technical Change. A Comparison of the Technological Impacts of Policy Instruments*. Cheltenham: Edward Elgar.

14 See World Commission on Environment and Development (1987) *Our Common Future*. Oxford: Oxford University Press.

15 For this argument see Irwin, Alan (1995) *Citizen Science. The Study of People, Expertise and Sustainable Development*. London: Routledge.

Chapter 2

Nurtured spaces

The future presented at autosalons and in numerous automobile magazines includes the enticing presentation of environmentally benign vehicles. Examples are electric vehicles powered by fuel cells, hybrid-electric vehicles with small petrol or diesel engines generating electricity on-board, natural gas vehicles, lightweight vehicles built with composite materials instead of metal, and vehicles for public individual transport systems.[1] Only very few of these vehicles ever become available for sale and if they do, it is often years after they appeared at a show or in a magazine.

This raises the question why such technologies are not introduced into the market place when their benefits to society are so evident. Is there no market for them? This is what the automobile manufacturers tell us. But why is there no market? Is it because consumers do not want to incur extra costs for environmental benefits? Or are the reasons political, namely the failure of policy-makers to make environmental benefits an integral part of the structure of incentives and constraints in which people trade and interact? Is it perhaps that manufacturers *think* that there is no market, or do they consider the market for environmentally desirable automobiles less attractive than the market for gasoline automobiles? As we will argue, there is not just one barrier to the introduction of alternative vehicles, but a whole range of factors that work against their adoption. These factors are interrelated, making a policy approach aimed at reducing individual barriers one by one less likely to succeed.

In the following section, we will take a closer look at the different factors that affect the development and use of new transport technologies, in particular how these factors impede a shift to more sustainable transport technologies. These barriers will be discussed individually although it is the combined occurrence of the barriers that is responsible for the slow transition to more sustainable transport technologies. The following sections pursue theoretical issues in innovation studies, drawing on the history and sociology of technology, and evolutionary economics. This theoretical review includes a discussion of concepts such as technological regime, niches and niche branching, and regime-shifts. It will lead to the proposal for a new kind of tool that facilitates the nurturing of spaces for implementing technological options that hold a promise for sustainable development. This tool is Strategic Niche Management.

Why is there under-utilization of sustainable technologies in transport?

Technological factors[2]

One important barrier to the introduction and use of new technology is that new technologies often do not fit well into existing transportation systems. The use of the new technology may require complementary technologies that are perhaps in short supply or expensive to use. The introduction of battery-powered electric vehicles, for example, will require the development of an infrastructure for charging batteries. It may also be that the basic technology itself needs to be further developed. In the early phase of their development new technologies often prematurely reflect user needs and are too expensive because of small-scale production. They need to be optimized. A related factor is that the new technologies have not yet been tested by consumers on a large scale. Actual large-scale use will stimulate an iterative process of redesign and lead to new, unforeseen design specifications.

These technological barriers have been given increased attention over the last couple of years, especially in connection with various experiments with new technologies (electric vehicles, natural gas vehicles, etc.).

Government policy and regulatory framework

Government policy may also be a barrier. Even though governments are committed to environmental protection and other social goals, often they are not putting out a clear message that there is a need for specific new technologies. In a sense, the signals are confounded because nearly all new technologies are stimulated by R&D subsidies even though it is not clear what specific role each should play in a future transportation system. Elzen *et al.*, for instance, in their international study of technology policies for transportation found no clear statements to guide developers, planners and investors towards sustainable development.[3] The manufacturers therefore remain uncertain about the market developments and will be reluctant to invest in risky alternatives.

Moreover, the existing regulatory framework may actually form a barrier to the development of new technologies. For instance, the very strict safety requirements in the Japanese law on natural gas drove up the price of on-board gas cylinders and refuelling stations to five times the level of other countries, until the laws were changed in the mid 1990s. The California Zero Emission Vehicle (ZEV) legislation has strongly stimulated the development of electric vehicles but for a long time discouraged the development of hybrid-electric vehicles, although the latter may be cleaner if the emissions by electricity production plants are taken into account.[4] Legislative flexibility to accommodate new technologies is often inadequate, partly because some of the actors may oppose the innovations in question.

Cultural and psychological factors

There may also be cultural and psychological barriers. In this century, the automobile, with its high speed and its image of freedom of the road at any given time, has become an icon of the modern lifestyle. Values such as flexibility and freedom are associated with the possession and use of a car. For many automobile users, owning and driving a car is a way of expressing individual and societal identity: the car is a status symbol.

Car manufacturers as well as consumers and car salesmen have an idea of what a car is and what it should be able to do. Alternative transportation technologies will convey different images and represent different values. Unfamiliarity with the alternatives often leads to scepticism because all actors judge the new technology based on the characteristics of the dominant technology. An example is the so-called idle-off device that was first offered on some Volkswagen models. This device shuts off the engine when the car is stationary or slowing down. This feature can cut fuel consumption in the city by 20–30% and also strongly reduce emissions. When the car accelerates the engine will restart automatically. The idle-off device has not become a success, because Volkswagen and the dealers do not dare to promote it. They think that the drivers will fear that the engine will not restart and therefore prefer the certainty of hearing the engine run even when the vehicle is stationary.[5]

Demand factors

There are also economic barriers, which have to do with prospective users' preferences, risk aversion and willingness to pay. The new technologies have not yet proved their value, so consumers are not sure what to expect. The meaning and implications of the new technologies have yet to be specified by their application in practice. Also new technologies may also not meet the specific demands of consumers, which means that changes in demands and preferences may be required in order to introduce these technologies. The battery-powered electric vehicle's limited range will force its user to adapt his or her travel patterns. Only a few consumers will accept lesser performance in return for lesser environmental impact. Manufacturers can easily justify not introducing a new transportation technology by pointing to consumer fears and insecurity. The vehicle market, for instance, is very sensitive, and a loss of market share because of the failure of a new product may cause serious problems. The manufacturers of existing technologies prefer to avoid risks by building on current consumer preferences. In addition, the automobile dealers are reluctant to promote cars to consumers that do not meet traditional consumer preferences.

Most manufacturers have spent many years and untold millions building brand equity, and are sceptical that consumer demands can be changed. Accordingly, industry argues that it cannot manufacture products for which there is no clearly articulated consumer demand. Taken to the extreme this logic suggests that the auto industry would never be able to launch a new product successfully. But witness the success story of the minivan in the United States: consumer research in the late 1970s had indicated a widespread sentiment in favour of a small people-mover van in the United States. The American car manufacturers began working on such a van, but Ford concluded that the vehicle would become too costly and General Motors considered the market too fragmented.[6] Only Chrysler went ahead in an all-or-nothing gamble in the face of bankruptcy, and hit instant success. Ford and GM then followed. The U.S. minivan market currently comprises unit sales of over one million vehicles. This example is not directly comparable to the proposed market for environmentally benign vehicles; minivan buyers did not have to settle for less with regard to comfort and performance, whereas battery-powered electric vehicles have limited range and speed, and recharging the battery is very time-consuming.[7] Nevertheless, the case of the minivan shows that even in seemingly saturated markets, manufacturers can use new products and configurations to tap previously overlooked pockets of consumer demand.

Another important demand factor is the price of the product. New technologies are often expensive due to the small absolute scale of manufacture and because they have not benefited from dynamic learning economies of production.[8] The high price that results from the high unit costs of production is a significant disadvantage in the competitive automobile market. Even relatively simple new technologies (such as, for example, the pre-heated catalyst) have a hard time because they increase the cost price.[9] However, Toyota's launch of the Prius hybrid-electric car in Japan shows that a manufacturer willing to incur temporary losses in order to build up market presence can be quite successful. Toyota's strategy to sell the Prius well below the break-even price paid off when enough cars had been sold to reduce the break-even price to the actual price.

Production factors

There are also barriers on the supply side. The development from prototype to mass product is lengthy and cumbersome, but above all it is risky. There may be a chance to develop a new market, but the incentive for the automobile industry to introduce a product onto the market is not high when it is far from certain that the consumer is interested in buying it or when there are no external factors, such as legislation, compelling the manufacturers to act. Investing in new technologies may mean that the sunk costs in existing production facilities will never be recouped. Moreover, incumbents are wary of entering markets where their existing capabilities and routines – their core competencies – would be inapplicable.

To the automobile industry, the mass production of cars with combustion engines is just such a core competence. Industry organization is aligned to this competence: technically (in terms of products, production processes and R&D activities); organizationally (in terms of modes of control, marketing and strategies); and culturally (in terms of recruitment promotion and retention policies). Generally, enterprises may target their production strategies at: (1) *cost leadership* – offering products at the lowest price on the market; (2) *differentiation* – offering exclusive products (for example of a specific brand) for a large market; or (3) producing for *market niches* – producing a limited assortment for a specific group of customers.[10] The major car manufacturers predominantly choose strategy (1) or (2); they have limited or no competence in producing cars for market niches. Therefore the manufacturers are only interested in alternative vehicles when these can be produced for a large market.

In such a situation it often takes new ventures to market new products. These do not stand much of a chance, however, if they are not backed by sufficient capital. This creates an additional problem since banks are reluctant to invest in risky projects, and governments only grant subsidies for R&D and not for marketing a new product. Moreover, new entrants often lack the operational capability to produce cars of a constantly high quality. Taken together, these factors represent significant barriers to entry. Co-operation between newcomers and the existing car industry might be able to change this. A good example is the co-operation between Swiss company SMH and Mercedes. This joint venture eventually resulted in the launching of the 'Smart' two-seater car in 1998.[11] Examples of small companies encountering major financial problems include the Swedish company Clean Air Transport (CAT) and the American firm US Electricar. CAT missed an order for 10,000 hybrid vehicles (the prize in a competition organized by the city of Los Angeles) because Swedish financiers did not trust the company's capability to deliver. US Electricar was forced to abandon the low-profit production of conversion-electric vehicles because of shareholder dissatisfaction as reflected in the collapse in the price of its publicly traded shares.

There are, however, more successful examples of new enterprises such as the French S.E.E.R., Swiss TWIKE, and Norwegian PIVCO (which was bought by Ford in early 1999). For the moment, these companies are all small-scale manufacturers.

Infrastructure and maintenance

The introduction of new technologies may also require adaptation of infrastructure. New distribution systems may have to be established, as for hydrogen or methanol fuels, or special provisions may need to be made, for example for charging electric cars and refilling natural gas vehicles. Another adaptation concerns the maintenance requirements of new types of vehicles. Mechanics in garages must be introduced to new technologies in order to be able to check and repair the new vehicles. A characteristic of infrastructure and maintenance investment is its threshold value: only with a relatively high number of vehicles does it become profitable to create a new infrastructure, even though the vehicles require such an infrastructure from the very beginning. Crucial questions thus include: who is responsible for the development of the infrastructure and how can initial costs be defrayed? Another problem is the so-called sunk investments in existing infrastructure. The groups in charge of the current infrastructure constitute a strong lobby for their own interests.

Undesirable societal and environmental effects of new technologies

New technologies may solve certain problems but they may also introduce new ones. Electric vehicle batteries could cause additional waste problems; some alternative fuels lead to an increase of certain types of emissions; the growing of crops required for the production of bio-fuels requires large amounts of land, which are then not available for other purposes (growing food crops or preserving nature, for example); the availability of cheap and very economical vehicles may trigger a demand response in the form of increased vehicle-mileage. Significant research will be needed to determine how such problems can be solved. In the meantime, such problems affect the image and performance of the new technology. Finally, the promotion of a specific technological path is always associated with a certain risk of creating new 'lock-ins': good solutions from today's point of view may preclude the development of better solutions in the future.

There are different ways to deal with these problems. The traditional approach is to overcome the barriers through policies especially designed to deal with each one: information dissemination, the use of subsidies to reduce costs and stimulate demand; public investment in infrastructure; support for research to improve the technologies and reduce negative side-effects for the environment or human health. Such policies address barriers one by one; they do not seek to exploit synergies because they fail to see how the factors are interrelated, especially how technical aspects are related with social and institutional ones. Rather than a set of discrete barriers, we face a *structure* of interrelated factors that feed back upon one another, jointly giving rise to inertia in, and specific patterns of, technological change. But what exactly is this structure and how does it affect technological choices of developers and users? These questions will be examined in the next section.

The structured nature of technological change:
technological regimes and paradigms

The existence of patterns in technological change is widely recognized. Examples include miniaturization in micro-electronics, the use of information technology in manufacturing and offices, the electrification of products and processes, the use of piston engines in post-war aviation, the production of electricity and motive power through the combustion of fossil fuels, and, on the consumer side, the use of automobiles for transport (about 80% of land-based passenger transport is done by car). Economists, historians and sociologists have studied these regularities in technological change and have proposed concepts to account for ordering and structuring of technology. We will describe two concepts that have been highly influential in social studies of technology: the concept of technological regime used by Nelson and Winter, and Dosi's concept of technological paradigm.[12]

The concept of a technological regime was coined in the 1977 article 'In Search of Useful Theory of Innovation' by Nelson and Winter. In this article they noted that the problem-solving activities of engineers were not fine-tuned to changes in cost and demand conditions, but relatively stable, focused on particular problems and informed by certain notions of how these problems could be dealt with. Citing the example of the DC3 aircraft in the 1930s, which defined a particular technological regime (metal-skin, low-wing, piston-powered planes), they observe:

> Engineers had some strong notions regarding the potential of this regime. For more than two decades innovation in aircraft design essentially involved better exploitation of this potential; improving the engines, enlarging the planes, making them more efficient.[13]

Dosi introduced the idea of a technological paradigm, analogous to Kuhn's concept of a scientific paradigm. A technological paradigm consists of an exemplar (an artefact that is to be developed and improved) and a set of (search) heuristics, or engineering approaches, based on technicians' ideas and beliefs of where to go, what problems to solve, and what sort of knowledge to draw on.

The idea of a core technological framework that guides industrial research activities has gained wide recognition in modern innovation theory. An advantage of this approach is its connection with existing engineering ideas and approaches, which the traditional economic notion of a production function fails to do. But as an approach to explain socio-technical change it is limited, because it focuses too much on cognitive aspects of problem-solving activities and too little on the interplay between cognitive, economic and other social factors that force technological problem-solving in certain directions.

This interplay can be perceived as a co-evolutionary process of variation and selection, in which external selection pressures are anticipated by the innovator organization and incorporated into company R&D and production policies; the external selection environment in turn is shaped by the policies of the innovator vendor and a host of other actors who strive to promote (and control) a particular technology.[14] Engineering activities are embedded in larger technological regimes which not only consist of a set of opportunities but also of a structure of constraints in the form of established practices, supplier-user relationships and consumption patterns. Society's preference for the internal combustion engine thus depends not only on the prevailing interpretative framework of engineers, but also on the embeddedness of the combustion engine in engineering practices, production plants and organizational routines, and the embeddedness of automobiles with internal combustion engines in fuel distribution systems, travel and mobility patterns, and automobile repair and maintenance practices.

If we take the co-evolutionary dynamics of technical change as a starting point, we need a broader definition of technological regime. A technological regime needs to encompass both the paradigmatic framework of engineers and the system elements of a technology. The definition of technological regime we use is:

> the whole complex of scientific knowledge, engineering practices, production process technologies, product characteristics, skills and procedures, established user needs, regulatory requirements, institutions and infrastructures.[15]

A technological regime incorporates a cognitive and normative framework and a set of (functional) relationships between technology components and actors throughout the product chain, which forms the basis for individual and collective action. A technological regime is the context for technological and economic practices within a product chain, which prestructures the problem-solving activities that engineers are likely to undertake and the strategic choices of companies.

The term regime is used rather than paradigm or system because it refers to *rules*[16] – not just rules in the form of a set of commands and requirements, but also rules in the sense of roles and established practices that are not easily dissolved. These rules consist of search heuristics of engineers, product standards, manufacturing practices, standards of use, and the division of roles. Like a political regime or a regulatory regime, a technological regime implies a set of rules. These rules guide (but do not fix) the kind of research activities that companies within a technology system are likely to undertake, the solutions that will be chosen and the strategies of actors (manufacturers but also suppliers, government and users).[17] The idea behind the technological regime is that the existing complex of a technology extended in social life imposes a grammar or logic for socio-technical change, the same way as the tax regime or the regulatory regime imposes a logic on economic activities and social behaviour. Our definition is thus more in line with the way in which the term regime is used in political science and policy studies.

Technological regimes, in the way we use the term, are a broader, socially embedded version of technological paradigms. In our view, the focused nature of socio-technical change is accounted for in large part by the *embeddedness* of existing technologies in broader technical systems, in production practices and routines, consumption patterns, engineering and management belief systems, and cultural values much more than it is by engineering imagination. This embeddedness creates economic, technological, cognitive and social barriers for new technologies.

This notion of technological regime helps to explain why most change is of the non-radical type, aimed at regime optimization rather than regime transformation. It helps to explain why so many new technologies remain on the shelf, especially systemic technologies with long development times that require changes in the selection environment (in regulation, consumer preferences, infrastructure, and price structure). Radically new technologies require changes both on the supply and demand side, which usually take time and meet resistance, even inside the organization in which they are produced. Firms vested in the old technologies will be more inclined to reformulate their existing products than to do something radically new that may involve great risk to the firm. There is also an incentive for outsiders to develop innovations that can be easily integrated into existing processes and products. This is not to say that technological change is all simply a matter of economics. As noted by Rosenberg and Fransman, firms have a restricted technological horizon and a bounded vision, which serve to focus their exploratory activities upon problems posed by the existing product.[18] This means that there are path dependencies that act to contain radically new technologies. Both supply-side and demand-side changes are needed to introduce radically new technologies successfully. These changes consist of new ideas, production and user practices, the development of complementary assets and institutional change at the level of organizations and markets. These findings are confirmed by historical studies of technological transitions.

Dynamics of regime-shifts

In this book we are interested in technological regime changes rather than in conservative responses within the existing regime. This raises the question as to what is involved in changes of technological regime. The history of technology does not provide a definite answer to this question. Obviously, each technological transition or regime shift is unique in its own way, but there are also general features. In a study of technological transitions the following elements were identified as key aspects of technological regime shifts.[19] These are:

1   Long periods of time. It often takes 50 years for a new technology system or regime to replace an old one.

2   Deep interrelations between technological progress and the social and managerial environment in which they are put to use. Radically new technologies give rise to specific managerial problems and new user-supplier-relationships; they require and lead to changes in the social fabric and often meet resistance from vested interests; moreover, they may give rise to public debates as to the efficacy and desirability of the new technology.

3   New technologies tend to involve 'systems' of related techniques; the economics of the processes thus depend on the costs of particular inputs and availability of complementary technologies. Technical change in such related areas may be of central importance to the viability of the new regime.

4   Perceptions and expectations of a new technology are of considerable importance. These include engineering ideas, management beliefs and expectations about the market potential, and, on the user side, perceptions of the technology. These beliefs and views of the new technology are highly subjective and will differ across communities. They are also in constant flux, and the progression of these ideas may be either a barrier or a catalyst to the development of a particular technology.

5   The importance of specialized applications in the early phase of technology development. In the early phase of a radically new technology there is usually little or no economic advantage of the technology; moreover, the existing technologies tend to improve during the development phase (the 'sailing ship' effect), rendering open market competition even more difficult.

Technological regime shifts thus entail a number of structural changes at different levels. The emergence of a new technological regime implies the simultaneous evolution of these changes. This simultaneous evolution is a co-evolutionary process: technological options, user preferences and needed institutional changes are not given *ex-ante*, but need to be created and shaped. Users, for example, do not have fixed demands that are fulfilled with a new technological option. The market (in the sense of observable demand) does not exist in cases of regime shifts. Instead, many case studies have shown how user demands are articulated and expressed in the process itself, in interaction with the technological options available. This process also works the other way round. Producers learn new ways to view their own technology. For example, at the beginning of this century, no articulated demand for automobiles existed. Producers gradually learned to distinguish the relevant product attributes, and drivers learned how to the use a car (i.e. for which purposes).[20]

Users often play an active role in this process. This is the case for professional users and user firms and also for end-users.[21] For example, when the telephone was introduced, end-users discovered sociability, using the telephone for chatting and other social purposes, while the industry tried to reduce this unnecessary and frivolous use. Industry sold the telephone service as a practical business and household tool, and it took some time before suppliers realized the market potential of social talk.[22] The example of the telephone also illustrates another important point. New technological regimes are not created; they evolve through the actions and strategies of many different actors. In this sense, regime-shifts gradually exceed the capability of any single actor to maintain control of the overall process of systemic change.

If a regime-shift is encompassing, complex, and has long lead-times, how then can we understand its emergence? Does a massive and complex change need a similar beginning? The answer is no, the start can in fact be very modest. Regime-shifts often start at the periphery of existing dominant technological regimes in small, isolated application domains. Kemp *et al.* have confirmed the importance of specialized applications in early phases of technology development. In addition, in a six-volume history of Dutch technology in the nineteenth century consisting of 24 case studies of various innovations covering many sectors in the Dutch economy, Schot concluded that technologies often first appear in niches.[23] In the historical literature many other examples can be found. The steam engine was developed by Newcomen to pump up water from mines. Clocks were first used in monasteries where life was arranged according to a strict timetable. The wheel was first used for ritual and ceremonial purposes. The telegraph was introduced by the railways, and the rapid press was first employed in the production of newspapers.[24] Here we will discuss one example in more detail to elaborate on the process of niche development: the emergence of gas light in the nineteenth century.[25]

A regime-shift through niche branching: the example of gaslight

The history of artificial light in the nineteenth century is mainly one of more light. But changes were not merely quantitative. By the end of the nineteenth century, light was produced with radically new methods; a new technological regime emerged. Gas lamps, incandescent lamps and paraffin lamps prevailed over candles and oil lamps and new norms for lighting were developed. How did the regime-shift take place? The first gas lamps were used in textile factories in England at the beginning of the nineteenth century. Textile manufacturers were interested in a cheap and fireproof alternative to candles and oil lamps, and they were willing to invest in an immature technology. Thus, their factories became the first experimental stage. They accepted the drawbacks of gas light (the heat and smell). So, the first experience with gaslight came from experiments of textile manufacturers, and a niche was formed.

A qualitative leap was made with the development of a second niche, that of street lighting. This required substantial technological development, especially the construction of a network of pipes to connect the central gas factory with a series of distributed lamps. The network was not only used for street lighting, but also for the illumination of factories (next to textile factories), theatres, coffee houses and other public spaces, so a series of niches emerged. The second (series of) niche(s) was located in London, but soon spread to other cities in England. After some time, a third market niche was opened up, that of wealthy private individuals. From England, gaslight made its way to continental Europe. The Imperial Continental Gas Association was established for this purpose. In a number of European countries there followed developments similar to those in England, spreading from small niches in specific places (factories and theatres) to street lighting and private homes.

By 1890, the new lighting regime had reached a stable equilibrium. However, it was also challenged by a new competitor: electricity. In this case, the example of gas helped establish electric lighting; the notion of a central facility linked to a system of distribution already existed and could be imitated and used (including institutions, regulation, views of user needs, etc.). An important issue in this new competition was that electricity solved a number of problems that had arisen with gaslight, notably heat and safety. However, electricity was also first applied in niches, such as public displays and local transportation (trams, trolley buses). When electricity was introduced, gaslight went through new developments. Many improvements were made, such as better pipes, the slot-meter, the gas mantle, and upgrading of the production process. In the end electric light prevailed over gaslight, but by then gas had found a new niche for cooking and heating.

What kind of patterns of niche development can be discerned in this short history of light? The first thing to observe is that gaslight emerged in the market niche of the textile industry. Due to distinct selection criteria in this niche, the textile industry was prepared to accept the disadvantages of gaslight (heat and smell). These drawbacks were less consequential in large spaces such as textile factories, where at the same time great importance was attached to the intensity of light and its relative fire safety. Clearly, gas lighting had specific advantages over the prevailing oil and candle technologies. In addition, this first niche provided the resources to sustain the innovation, that is time, energy, capabilities, knowledge, and finance for an alliance (network) able to produce and use the new technology. The alliance sheltered and developed the new technology. This was necessary, as the technology was still rather crude and not (yet) up to the demands of other potential markets.

From this first niche, a number of new niches developed, mainly for street lighting and public spaces. This process of *niche branching* includes the emergence of new application domains and the creation of a bandwagon effect (that is a wider diffusion) through replication of the niche elsewhere. London was imitated by other cities in England and on the continent. During these developments the technology was improved and further technical choices were made, partly in response to new selection criteria. For example, gas could be produced from coal or oil, and distribution could be via conduits or canisters. Rip has summarized the process of niche branching as follows:

> Technological change is not a continuous process along dimensions of increasing functionality. It is more like a patchwork quilt, or if one prefers a different metaphor, the way yeast cells grow. Developments branch off in different directions, cross-connections and interactions occur, and niches, that is limited and relatively easy or advantageous domains of application and further development, strongly determine what steps can be taken productively. The eventual shape of a technology, its usage and the way it is embedded in society can be very different after 5, 10 or more years than it looked at the beginning.[26]

Finally, we would like to point to another characteristic of this process of niche branching. Consumers' demand developed; people learned to appreciate gas lighting and its intensity. This led to a new reference point for evaluating the traditional alternatives. Suddenly, candles and oil lamps were perceived in a new 'light'; they looked pale. Not only consumers changed in the process; also cities learned gradually how to respond to new opportunities. Eventually, they took on a new identity: that of gas producer, a new role for municipalities and one which some resisted for a long time. Also, a new firm was created to promote gas light in continental Europe. In short, through a learning process, a new world, and a new technological regime emerged in which gaslight invaded the markets of other light sources. Life without gaslight became almost unimaginable. Gas lighting became the new standard; the process had become irreversible.

The nature of a niche

In an interesting article, Daniel Levinthal introduced a niche theory of technical change in order to account for two contrasting perspectives of technical change – one stresses gradual or incremental change, and the other emphasizes rapid and discontinuous shifts.[27] He bridges both perspectives by arguing that the critical event in technical change is not a transformation of technology, but the specialist application of existing technology to a new domain of application (niche). Subsequently, as a result of distinct selection criteria operating in the niche and the degree of resource abundance, a radical change might occur. While the initial *speciation* event might be minor in the sense that the technology does not differ substantially from its predecessor, it triggers a divergent evolutionary path.[28]

This theory fits the structure of the example presented above, but differs from it in three illuminating and crucial aspects. First, Levinthal emphasizes that the new technology is always commercially viable and profitable in the niche. Accordingly, a key element in niche development is the (Schumpeterian) entrepreneurial activity of linking a technology to an application domain. We want to stress that in these first niches commercial viability might well be absent. The first applications of electricity at world fairs, theatres and public events had symbolic value; they brought excitement. The first applications of aeroplanes and cars in races were never commercial successes; indeed, the motivation to engage in such activities was not primarily economic in nature.[29] Expectations play a crucial role in early phases of technical change.[30] Technologies in the making have yet to prove themselves (in terms of technical, social as well as commercial viability). Parties that apply a new technology, therefore, often construct and communicate positive expectations in order to make actors (including themselves) *believe* that it will yield returns in the future, but they cannot be certain of these returns.

Second, we do not locate the innovative power in a single isolated entrepreneur, but suggest that network development and alignment work of actors is crucial to exploit niche opportunities. Third, Levinthal's starting point is to create a bridge between incremental and radical change by connecting an incremental start to radical implications, what might be called niche drifting. For us, the distance between the niche technology and the dominant technology (we prefer to talk about technological regime, a concept which is missing in Levinthal theory) might vary and does not only consist in a difference in technological characteristics, but also in differences in users and their preferences and institutional set-up.

Nature of regime-shift

How can we explain regime-shifts? Is the direction and success of a regime-shift already determined in an earlier phase of technology competition by historical lock-in, as suggested by Brian Arthur (in his increasing return of adoption model)?[31] If one technology gets ahead by good fortune, it gains an advantage, attracts more adopters making the market tip in its favour and leading to a domination that is almost irreversible. Further increasing adoptions makes one technology more attractive because of learning effects, network externalities, and a better understanding of a specific technology, which in turn excludes other technologies. The process of technology introduction is, thus, self-reinforcing or path-dependent. Choices made for local and specific reasons and at specific times determine subsequent history. Given other circumstances, a different technology might have been favoured early on, and would have become the dominant technology. For this reason history matters.[32] The classic example is the QWERTY keyboard introduced in the United States at the end of the nineteenth century with the explicit aim of lowering typing speed, for otherwise the hammers of the typewriters would clash and jam. The risk is no longer real, but repeated attempts to introduce faster keyboards have failed.[33]

In the niche model, lock-in and path-dependency assumptions are relaxed. Various technological options can co-exist over a long period, precisely because of the existence of niches requiring other functionalities. In addition, the dominant technology (when out-competed by a new one) may continue to thrive in other niches, as the story of gaslight has shown. Niches also persist because actors such as firms and governments act strategically by keeping certain options alive which might be important for future competition or other broader societal goals.[34] Although technological diversity is not completely reduced, at the same time the emergence of technological regimes does direct research, development, use and regulation in specific directions leading to dominance of specific artefacts.[35]

How then can we explain the emergence of a technological regime in the midst of technological diversity? Factors summarized in the increasing return of the adoption model of Arthur and other evolutionary economists do play a major role here. However, the model is weak in explaining the emergence of niches and their success in an early phase when various niches compete against the background of an existing dominant regime. The model is reserved for situations where two or more alternatives appear at the same time, compete, and are equally positioned to create a bandwagon effect.[36] Thus, both for explaining early niche development and for explaining bandwagons, a more complex model is needed. The model we offer, based on work by Rip and Kemp and Schot, is a *multi-level* one that stresses developments at various levels which then could *couple* (create positive and negative feedbacks) at specific times.[37] This coupling is, in turn, a major source of bandwagon effects. Of course, this is an unplanned and contingent process, which cannot be precisely anticipated. Actors can, however, enhance the *probabilities* for coupling to occur, and this is precisely one of the underlying aims of strategic niche management. Our hypothesis is that a regime-shift requires three types of coupled developments:

- Processes of niche developments of novelties followed by increasing returns of adoption;

- Erosion of opportunities to make progress within the regime; and

- Emergence of new external opportunities and constraints which challenge the problem solving capability of the existing regime. Such external development can be events (such as a war or a scientific breakthrough that allows for new technical developments) or broad trends such as urbanization.

These developments play themselves out at various levels. Niche development is the local level of innovation processes. Of course, successful innovation implies that the innovation will become more broadly known and adopted, hence more global. The level of a regime is a more global level of shared understanding and rules that orient actors' behaviour at the local level. At some point actors might start to rethink the viability of technological options in the existing regime. A classic example from the history of technology is the development of aerodynamic theory in the 1920s, which suggested that the traditional regime of piston engine and propeller combination in aeroplanes would never be able to compete with turbojets. Consequently, the design constituency lost faith in incremental improvements to the existing regime. It is important to note that this process did not occur overnight, and was heavily influenced by World War II.[38]

The third level is also global, although its effects can differ widely at the local level. This level has been referred to as a socio-technical landscape and can be defined in two ways. First, the socio-technical landscape can be thought of as a set of connected technological and societal (hence socio-technical) trends, deep structures and major events that influence the opportunity structure for technologies embedded in regimes as well as new promising alternatives. These sets of factors are literally a landscape because they accommodate some developments more easily than others do. Second, the landscape is not influenced directly by the success of local innovation processes. So these factors influence the process but regime-shifts in specific sectors will not affect the landscape itself dramatically. Of course, if a number of regime-shifts occur, the landscape will be changed. For example, the emergence of electricity led to changes in factory regimes, transportation regimes, and household regimes and thus to a new kind of electrified society and economy. This is a special case of a pervasive technology.[39]

Looking again at the case of gaslight, we can explore how this process of coupling works. Industrialization and the emergence of an innovative textile industry were important for the first application. This application turned out to be a success, leading to a process of niche branching and competition with alternatives. The Napoleonic wars led to considerable increases in the price of oil lamps and candles, whereas the price of coal, the raw material for gas production, remained low. Urbanization created public safety problems in London and a need for more lighting and thus led to a reorientation and reassessment of oil lamps and candles. Both the wars and urbanization eventually were factors in the process of adoption of gaslight. More examples of coupling (positive and negative feedback) will be seen in Chapter 4 in our discussion of the history of the car. Here we will focus on what is happening inside the niche. Which factors make early niche development successful?

Successful niche development

We would like to introduce two measures for evaluating the success of early niche development: quality of learning and quality of institutional embedding. Learning refers to a range of processes through which actors articulate relevant technology, market and other properties. This is a learning process because outcomes are not known beforehand, but actors have to work hard to define outcomes. Learning concerns a number of aspects:

- Technical development and infrastructure: this includes learning about design specifications, required complementary technology and infrastructure;

- Development of user context: this includes learning about user characteristics, their requirements and the meanings they attach to a new technology and the barriers for use they encounter;

- Societal and environmental impact: this entails learning about safety, energy and environmental aspects of a new technology;

- Industrial development; this involves learning about the production and maintenance network needed to widen diffusion; and

- Government policy and regulatory framework: this involves learning about institutional structures and legislation, the government's role in the introduction process, and possible incentives to be provided by governments to stimulate adoptions.

Learning can be limited to first order learning. That is when, in the niche, various actors learn about how to improve the design, which features of the design are acceptable for users, and about ways of creating a set of policy incentives which accommodate adoption. However, for niche development to result in a regime-shift, another kind of learning process is needed – second order learning. In such learning processes, conceptions about technology, user demands, and regulations are not tested, but questioned and explored. Opportunity emerges for co-evolutionary dynamics, that is, mutual articulation and interaction of technological choices, demand and possible regulatory options. Co-evolutionary learning will also allow for what Brian Wynne has called collective value learning, that is clarifying and relating of various values of producers (designers), users and other third parties involved, such as governments.[40] Successful niche development consists of first order learning on a whole array of aspects, and the occurrence of second order learning.

The emergence of a new technological regime will change the selection environment for innovation. Through processes of niche development, this change will be prepared. This is the process of institutional embedding. Three crucial aspects of institutional embedding can be identified. First institutional embedding entrains complementary technologies and the necessary infrastructures, a factor that is also highly important for increasing return of adoption dynamics in later diffusion phases. Second, embedding produces widely shared, credible (supported by facts and demonstration successes) and specific expectations. Third, embedding enlists a broad array of actors aligned in support of the new regime. This network includes producers, users and third parties, especially government agencies. Numerous studies of product innovation and failure have shown that involvement of users is an important factor for successful market introduction, and lack of user involvement is a major cause of failure.[41] Particularly for innovations that serve broader societal goals, such as innovations leading to more sustainable transportation, third party involvement representing the interest of sustainable development is crucial. Alignment refers to a situation in which the actors have developed a stable set of relationships and can easily mobilize additional resources in their own organizations because the network is seen as an important, credible and strategic operation. In such situations, as Rip has suggested, so-called macro-actors, who have a separate responsibility to realize and maintain high alignment, are often available.[42] Thus, successful niche development assumes the development of complementary technologies, more robust expectations, and a broad, highly aligned network.

Defining Strategic Niche Management[43]

A main claim of this book is that Strategic Niche Management is a policy tool that can contribute to successful niche creation for new technological options. As argued in the introductory chapter, such options are abundant; however, they lack early diffusion. The innovation journey ends with prototyping. SNM is directed at overcoming this barrier to broader diffusion by exploiting niche dynamics. SNM aims to stimulate actual use. As several authors have argued, practical experience is necessary to generate the knowledge required to accommodate introduction. Such knowledge cannot be acquired in any other way, because the interactions between products and their use environments are too complex. For example, a study by Von Hippel and Tyre showed that a series of problems in two novel process machines could not be identified prior to field use for two reasons: existing problem-related information could not be identified in the midst of complexity, and machine users and others introduced new problem-related information after field introduction of the new machines. Designers can invest in fault anticipation strategies (such as simulation and scenario building) depending on costs and benefits, but learning by using is necessary. This implies the impossibility of designing a perfect technology in house. It needs to be tested and tried out in practice. There is, therefore, always a need for an experimental introduction of novel technologies into use environments with the intent to learn.[44]

These niches are market niches where a novel technology has specific advantages over the established technology. Both producers and users recognize this potential. Both the new technology and users' preferences need articulation that could in turn lead to new, yet unrecognized advantages, but the main actors involved accept that the new technology suits specific needs of the application domain. Yet, the technology might need specific protection measures, such as preferential treatment in the tax system or subsidies for reducing costs. In market niches regular market transactions prevail, that is producers and users negotiate about implementation and adoption. For some promising technologies, such market niches do not emerge spontaneously. In such cases, proto-market niches might be created, in what we call technological niches. In such technological niches specific advantages can be promised, but they are uncertain and not yet shared among the actors promoting the niches. Often, niche activities are geared towards identifying and testing assumptions about specific advantages. Technological niches come about in the form of experiments, and pilot and demonstration projects.

SNM refers to the managing of the process of (technological and/or market) niche creation, development and breakdown to enable regime-shifts. As we have shown, successful niche development requires specific learning modes and institutional embedding patterns. Subsequently, SNM policies have to be directed at creating such learning and embedding. SNM aims at broad and deep learning (second order learning), at the development of a broad aligned network, at widely shared specific expectations and, finally, at complementary technologies and infrastructures.

As to outcomes of SNM (and the further development of niches), four distinct possibilities can be distinguished:

1   Technological niche remains a technological niche through the set-up of
     follow-up experiments.   This might involve branching to a new
     application domain and replication in similar domains. Technological
     niche gestation might lead to expansion and upscaling of the niche.

2   Technological niche becomes a market niche. New experiments are not
     necessary any longer, but users start to recognize the advantages of the
     novel technology and suppliers are willing to invest in production on a
     small scale.

3   Market niche is expanding and branching in new directions leading to
     the emergence of new market niches.

4   Technological or market niche extinction. The novel technology fails to
     attract further support and becomes (again) an R&D option (albeit, this
     time less promising). Niche extinction does not imply that investments
     are lost. Many spillovers in terms of network development, technical
     learning and reputation gains can justify the risk of having tried. In
     addition, learning that a certain technology development is not desirable
     is also part of SNM.

The use of SNM as a policy tool does not lead automatically to a regime-
shift. Our multi-level model suggests the relevance of a number of other tools
as well (see Chapter 6). However it is a crucial stepping stone for the
development of viable alternatives. The existence of such alternatives will first
of all put pressure on the dominant regime to exploit all available options to
reduce environmental effects, and if niche development becomes coupled to a
set of wider changes it might contribute to a shift. Wider changes could, for
example, be a change of perception of remaining technological opportunities
in the dominant regime and the introduction of strict regulation which the
regime has difficulty meeting. For encouraging such a shift, governments need
to turn to other kinds of policies. In Chapter 6, we will compare SNM with
other possible policies.

If SNM wants to maximize chances for a regime-shift, it is important to
pick the right kind of technologies. SNM is strategic, because it does not pick
'winners' in the sense of short-term attainment of market share, but those
technologies which have potential for the longer-term process of regime-shifts.
These technologies should allow for development not too far away, but at the
same time have probability for positive coupling with other levels. Potential
for regime-shift can be measured in terms of involvement of outsiders who do
not share regime rules, and acceptance of a new kind of rules as this offers
possibilities for co-evolutionary effects. For this reason, the first step in any
Strategic Niche Management process is a regime analysis. To this we now
turn.

Notes

1   Such systems consist of a fleet of vehicles that can be rented for short periods and central management that controls their location and disposition. Examples are the Praxitèle and TULIP systems currently experimented with in France.

2   The discussion of these barriers and included empirical examples are taken from Elzen, B., Hoogma, R. and Schot, J.(1996) *Mobiliteit met Toekomst. Naar een Vraaggericht Technologiebeleid.* Report for the Dutch Ministry of Traffic and Transport. The Hague: Ministerie van Verkeer en Waterstaat. For similar accounts of barriers we refer to Tengström, E. (1994) Why have the Political Decision-makers Failed to Solve the Problems of Car Traffic? Reports in Human Technology no. 2, University of Goteborg; and Hård, M. and Knie, A. (1993) The ruler of the game: the defining power of the standard automobile, in *The Car and Its Environments. The Past, Present and Future of the Motorcar in Europe.* Proceedings from the COST A4 Workshop in Trondheim, Norway, pp. 137–158.

3   Elzen, Hoogma and Schot, *op. cit.*

4   The ZEV legislation has meanwhile been changed to include hybrid-electric vehicles.

5   Canzler, W. and Knie, A. (1994) *Das Ende des Automobiles, Fakten und Trends zum Umbau der Autogeschellschaft.* Heidelberg: Verlag C.F. Müller.

6   Porac, Joseph F., Rosa, José A. and Saxon, Michael S. (1997) America's Family Vehicle: The Minivan Market as an Enacted Conceptual System. Paper for the Multidisciplinary International Workshop on Path Creation and Dependence, Copenhagen Business School.

7   The typical range of the latest generation electric vehicle is up to 200 km, the top speed 100 km/h, and charging may take some 6 hours, although 80% of battery charge can be restored within an hour through quick 'opportunity' charging. On the other hand, hybrid-electric cars have none of these disadvantages. Their market introduction could thus well be as successful as that of the minivans. Late 2001, Toyota had sold 70,000 Prius hybrid cars since the launch in Japan in December 1997; this includes 15,000 hybrids sold in the U.S. The car is priced in the same range as high-end market family-size sedans.

8   For this notion see Kemp, R. and Soete, L. (1992) The greening of technological progress: an evolutionary perspective. *Futures*, **24**, pp. 437–457.

9   Since an exhaust gas catalyst does not function well at low temperatures, pre-heating the catalyst markedly reduces the level of harmful emissions. The catalyst can be pre-heated with electricity from the grid or with a device that stores engine heat and releases it when the engine is started. Petrol cars can be equipped with a pre-heated catalyst without further changes to the vehicle. Alternatively, cold-start emissions may be stored and treated when the catalyst is warmed up.

10 Porter, Michael E. (1980) *Competitive Strategy; Techniques for Analyzing Industries and Competitors.* New York: Free Press.

11 The history of this cooperation and the role of outsider initiatives in the re-invention of the car are analysed in Truffer, B. and Dürrenberger, G. (1997) Outsider initiatives in the reconstruction of the car: the case of lightweight vehicle milieus in Switzerland. *Science, Technology and Human Values*, **22**(2), pp. 207–234. The cooperation ended in 1998 when DaimlerChrysler took full control of the company selling the 'Smart', after a conflict with the former outsiders over whether or not to launch a hybrid version of the car.

12 Nelson, Richard R. and Winter, Sidney G. (1977) In search of a useful theory of innovation. *Research Policy*, **6**, pp. 36–76; Dosi, Giovanni (1982) Technological paradigms and technological trajectories: a suggested interpretation of the determinants and directions of technical change. *Research Policy*, **11**, pp. 147–162.

13 Nelson and Winter, *op. cit.*, p. 57.

14 The selection environment is also shaped by the experience of users and the adjustment of users (both companies and consumers) to particular technologies. For a discussion of co-evolution of technology and society see Rip, A. and Kemp, R. (1998) Towards a theory of sociotechnical change, in Rayner, S. and. Majone, E.L (eds.) *Human Choice and Climate Change.* Columbus,Ohio: Batelle Press, pp. 327–399; and Molina, A.H.(1993) In search of insights in the generation of techno-economic trends: micro- and macro-constituencies in the

microprocessor industry. *Research Policy*, 22, pp. 479–506. Molina does not refer to the concept of co-evolution, but argues in a similar way (p. 483): '*Sociotechnical constituencies may be defined as dynamic ensembles of technical constituents (tools, machines, etc.) and social constituents (people and their values, interest groups, etc.), which interact and shape each other in the course of the creation, production and diffusion of specific technologies. Thus, the term "sociotechnical consituencies" emphasizes the idea of interrelatedness. It makes it possible to think of technical constituents and social constituents stressing the point that in the technological process both kinds of constituents merge into each other.*' We also refer to Garud, R. and Rappa, M.A. (1994) A socio-cognitive model of technological evolution: the case of cochlear implants. *Organization Science*, 5, pp. 344–362. In some other publications, instead of co-evolution, the wording quasi-evolution has been used. There a more extensive discussion on the way variation and selection processes relate can be found, see for example, Schot, J. (1998) The usefulness of evolutionary models for explaining innovation. The case of the Netherlands in the nineteenth century. *History and Technology*, **14**, pp. 173–200.

15 Rip and Kemp have accentuated the structured nature of a technological regime by defining a technological regime as the *coherent* complex of scientific knowledges, engineering practices, production process technologies, product characteristics, skills and procedures, and institutions and infrastructures that are labelled in terms of a certain technology (for example, a computer), mode of work organization (for example, the Fordist system of mass production), or key input (like steel or hydrocarbons). Since the accommodation between the elements in the complex is never perfect, it is perhaps better to talk about a *semi*-coherent complex. For similar definitions see Kemp, R. (1994) Technology and the transition to environmental sustainability. The problem of technological regime shifts. *Futures*, 26, 1994, pp. 1023–1046 and Kemp, R. (1997) *Environmental Policy and Technical Change. A Comparison of the Technological Impact of Policy Instruments.* Cheltenham: Edward Elgar.

16 Large technical systems as defined by Thomas Hughes can be seen as a special kind of regime, one in which material connections and the building up of an infrastructure are crucial to its diffusion. This creates special effects, such as the importance of load management, and leads to what Hughes has called momentum (the tendency of such systems continually to expand, often at the expense of alternative technologies). Hughes, Th. P. (1983) *Networks of Power. Electrification in Western Society 1880–1930.* Baltimore: Johns Hopkins University Press.

17 It is important to note that a technological regime does not totally fix technological choices. It is open to various kinds of change at the level of regime components and even the overall architecture. Technological regimes change in conjunction with the evolution of social needs, technological possibilities and organizational change such as new management systems.

18 See Rosenberg, N. (1976) The direction of technological change: inducement mechanisms and focussing devices, in his book *Perspectives on Technology.* Cambridge: Cambridge University Press, pp. 108–125; and Fransman, M. (1990) *The Market and Beyond. Cooperation and Competition in Information Technology in the Japanese System.* Cambridge: Cambridge University Press.

19 Kemp, R. *et al.* (1994) *Technology and the Transition to Environmental Stability: Continuity and Change in Complex Technological Systems.* Final report for SEER research programme of the Commission of the European Communities. Maastricht/The Netherlands: Maastricht Economic Research on Innovation and Technology.

20 Abernathy, W.J., Clark, K.B. and Kantrow, A.M. (1983) *Industrial Renaissance. Producing a Competitive Future for America.* New York: Basic Books, pp. 25–26. See also Green, Kenneth (1992) Creating demand for biotechnology: shaping technologies and markets, in Coombs, Saviotti, Rod, Paolo and Walsh, Vivien (eds.) *Technological Change and Company Strategies.* London: Harcourt Brace Jovanovich, pp. 164–184.

21 For the innovative role of user firms, see Hippel, E. von (1976) The dominant role of users in the scientific instruments industry. *Research Policy*, 5, pp. 212–239; see also Hippel, E. von (1988) *The Sources of Innovation.* Oxford: Oxford University Press.

22 Fischer, Claude S. (1992) *America Calling. A Social History of the Telephone to 1940.* Berkeley, CA: University of California Press, especially chapter 3.

23 Schot, J.W. (1995) Innoveren in Nederland, in. Lintsen, H.W *et al.* (eds.) *Geschiedenis van de Techniek in Nederland. De wording van een moderne samenleving 1800–1890*, Volume VI. Zutphen: Walburg Pers, pp. 217–240. An adapted version of this chapter has been published in English, see Schot, J., *op. cit.*

24 For these examples see Mumford, Lewis (1963, 1934) *Technics and Civilisation.* San Diego: Harcourt Brace Jovanovich Publishers, p. 12–17, and Bassala, G. (1988) *The Evolution of Technology.* Cambridge: Cambridge University Press, pp. 8–9.

25 Taken from Schot, *op. cit.*

26 Rip, Arie (1995) Introduction of new technology: making use of recent insights from sociology and economics of technology. *Technology Analysis and Strategic Management,* 7(4), pp. 417–431; the quote is on p. 418.

27 Levinthal, D.A. (1998) The slow pace of rapid technological change: gradualism and punctuation in technological change. *Industrial and Corporate Change,* 7(2), pp. 217–247.

28 This draws on the punctuated equilibrium framework of evolutionary biology, suggested by Gould, S. and Elridge, N. (1977) Punctuated equilibria: the tempo and mode of evolution reconsidered. *Paleobiology,* 3, pp. 115–151.

29 See Nye, David E. (1990) *Electrifying America. Social Meanings of New Technology.* Cambridge, Mass: MIT Press. He writes: '*Just as electric light for too long has been understood merely in terms of its practicality, so too electric traction has been studied primarily as a form of transportation, without recognizing that it altered the city's image, and together with spectacular lighting was involved in turning the urban landscape into a spectacle. A machine's social reality is constructed, and emerges not only through its use as a functional device, but also through its being experienced as a part of many human situations which collectively define its meaning.*' (p. 85)

30 See Lente, Harro van (1993) *Promising Technology. The Dynamics of Expectations in Technological Developments.* Unpublished PhD thesis, University of Twente.

31 Arthur, W. Brian (1988) Competing technologies: an overview, in Dosi, G. *et al.* (eds.) *Technical Change and Economic Theory.* London: Pinter Publishers, pp. 590–607.

32 See Dosi, *op.cit.*, and North, D.C. (1990) *Institutions, Institutional Change and Economic Performance,* Cambridge: Cambridge University Press.

33 For a summary of this example and references, see Utterback, James U. (1994) *Mastering the Dynamics of Innovation.* Boston, Mass.: Harvard Business School Press, pp. 5–7.

34 For a similar argument to relax the notion of path-dependency, see Islas, Jorge (1997) Getting round the lock-in in electricity generating systems: the example of the gas turbine. *Research Policy,* 26, pp. 49–66.

35 Windrum, Paul and Birchenhall, Chris (1998) Is product life cycle theory a special case? Dominant designs and the emergence of market niches through co-evolutionary learning. *Structural Change and Economic Dynamics,* 9, pp. 109–134. They have argued that the emergence of a dominant design is a special case. This might be true, but it is a widely applicable special case, see Utterback, *op. cit.*, for many examples.

36 See Islas, *op. cit.*, p. 51.

37 Rip and Kemp, *op. cit.* and Schot *op. cit.*, p. 191. See also Schot, J., Lintsen, H. and Rip, A. (eds.) (1998) *Techniek in Nederland in de Twintigste Eeuw,* Vol. I. Zutphen: Walburg, chapter 2, pp. 37–39.

38 Constant, E.W. II (1980) *The Origins of the Turbojet Revolution.* Baltimore: Johns Hopkins University Press.

39 For the effects of the introduction of electricity in all those areas see Nye *op. cit.* For an elaboration of pervasive technologies see Freeman, C. and Perez, C. (1982) Structural crisis of adjustment: Business cycles and investment behaviour, in Dosi, *op. cit.*, pp. 38–66. They distinguish four types of innovations: incremental innovations, radical innovations, changes of technology system (this is what we call a regime-shift) and changes in techno-economic paradigm (following from pervasive technologies)

40 Wynne, B. (1995) Technology assessment and reflexive social learning: observations from the risk field, in Rip, A., Misa, Th. J. and Schot, J. (eds.) *Managing Technology in Society. The*

Approach of Constructive Technology Assessment. London: Pinter Publishers, pp. 19–36.

41 For an overview see Leonard, Dorothy (1995) *Wellsprings of Knowledge. Building and Sustaining the Sources of Innovation*. Boston, Mass: Harvard Business School Press, especially chapter 7.

42 Rip (1995), *op. cit.*, pp. 426–427.

43 SNM was first suggested by Rip, A. (1992) Between innovation and evaluation: sociology of technology applied to technology policy and technology assessment. *Riviasta di Studi Epistemologici e Sociali sulla Scienza e la Tecnologia*, 2, pp. 39–68 as an R&D strategy. In Schot, J., Hoogma, R. and Elzen, B. (1994) Strategies for shifting technological systems: the case of the automobile system. *Futures*, 26(10), pp. 1060–1076 it has been developed into a strategy for technology introduction. Further development is visible in Kemp, R., Schot, J. and Hoogma, R., *op. cit.*, and Weber, M., Hoogma, R., Lane, B. and Schot, J. (1999) *Experimenting with Sustainable Transport Innovations. A Workbook for Strategic Niche Management*. Enschede/Sevilla: University of Twente/IPTS.

44 Hippel, E. Von and Tyre, E.M. (1995) How learning by doing is done: problem identification in novel process equipment. *Research Policy*, **24**, pp. 1–12; see also Leonard Barton, D. (1990) Implementation as mutual adaptation of technology and organization. *Research Policy*, **17**, pp. 251–267.

Chapter 3

Promises for sustainable transport

There are many options for making the existing transport regime of passenger transport more sustainable and these may be divided into two categories: those that contribute to *regime optimization* and those that contribute to *regime-shifts*. In many cases it is hard to establish beforehand in which category an option or innovation will fall. For some innovations, it is clear that they fit the existing regime, and thus will contribute to the ongoing process of regime optimization. Others may in principle support both optimization and regime-shifting at the same time: depending on the context in which they are introduced they may reinforce the regime or support a regime-shift. It is impossible to select innovations that, without more ado, will lead to a regime-shift ('to pick winners'), but for every innovation we can establish whether there is a *potential* for regime-shift. This entails making an assessment, for each specific technology, of the likelihood that it could lead to co-evolutionary dynamics. The questions are thus which innovations in the transport domain have the potential of contributing to a regime-shift, and what kind of shift would that be? In order to answer these questions, we need first to make an analysis of the present regime.

Rudimentary analysis of land–based passenger transport regime

Transport technologies and practices are part of specific regimes. We distinguish between a passenger transport regime with the two (sub)regimes of individual car-based transport and public transport and a freight transport regime.[1] We will not consider the latter in this book. The regime of individual car-based transport dominates in terms of satisfying mobility transport needs. It is based on the use of privately owned,[2] self-driven (i.e., not externally controlled), all-purpose vehicles, produced mainly from steel by large car manufacturers using mass production methods[3] and sold by automobile dealers that are linked to the manufacturers. All-purpose cars are vehicles that offer satisfactory performance over a wide range of dimensions, in particular: speed, driving range, acceleration, fuel consumption, safety, maintenance, and storage space. There are notable differences, however, even within price categories, especially with respect to fuel consumption and storage space.[4] Besides these privately-owned, all-purpose-cars, there are also more specific purpose vehicles often operated in fleets, including vans, buses, trucks and special vehicles such as roadsweeping machines. The use of all such vehicles is embedded in interorganizational relationships (between automobile and oil companies, insurance and leasing companies, transport departments, etc.) and a regulatory and fiscal system that gives the regime a measure of coherence and stability. Automobiles are also embedded in social and cultural systems, which means that a car is more than a convenient means of transport: it is a symbol of freedom, it signals social status and it is an expression of one's identity (*'you are what you drive'*).[5]

The regime of private car-based passenger transport is supplemented by a public transport regime of trains, buses, trams and metros, organized, controlled and operated by public transport companies. Public transport does not provide door-to-door transport services but uses fixed routes and schedules. In order to get to one's destination one may need to change transport modes at fixed transport nodes. In relative terms, car owners decreasingly use public transport. Only in large cities is public transport sometimes used more often than private cars. In Japan in 1991, nationwide the share of trains for commuting was below 30%; but was 50% in Osaka and over 70% in Tokyo.[6] Integration of the regimes of individual and public transport is not yet well developed, although 'intermodality' has become a key objective in transport policies in the last decade.

Both regimes of surface-based passenger transport are prone to specific problems for the individual user and society. The main problems in the case of the car are increasing congestion and parking problems, high energy consumption, extensive land use and pollution (including time pollution). Critics emphasize that the social costs associated with these issues are not reflected in the low direct costs of automobile use.[7] While the problems caused by the passenger car are a result of its success, the lack of success of public transport prevents it from making a positive contribution to solving these problems. Public transport takes up less space and, per passenger, consumes less energy and pollutes less, but it has difficulty attracting passengers due to long travelling times, lack of flexibility and (for men) the lack of privacy.[8] In Europe these problems have been aggravated recently by gradual cuts in government subsidies. There are local variations in the regimes, resulting from different industrial structures, government policies, climate and geography, cultural preferences etc., and the regimes also change over time. This is important, because it implies that there may be locations and circumstances where and when the regime is more receptive to technologies being developed outside it – in other words, when a niche for new technologies may be more easily created.

Table 3.1 characterizes the two regimes of land-based passenger transport in terms of key actors on the supply side – the system of production and

Table 3.1 The present situation: two highly disjunct regimes of land-based passenger transport

	The regime of private car-based transport	The regime of collective transport
Key actors at supply side	Oligopolistic car manufacturers who sell their vehicles through dealers, using intensive marketing via mass media to invest vehicle types with cultural values	Local, regional and national public and private transport companies who are in close contact with bus and rolling-stock manufacturers
Production system	High investment costs and technical and organizational complexities in car manufacturing that create an entry barrier for newcomers and for alternative vehicles	High investment costs, technical and organizational complexities in bus and rolling-stock manufacturing; lower entry barriers for alternative vehicles
Maintenance	External maintenance in garages by mechanics trained in the repair of internal combustion engine vehicles	Internal maintenance by public transport companies
Supportive infrastructures and policies	Car use is supported by extended road network and filling system and by land-use planning, fiscal policies, car insurance, leasing facilities and the partial internalization of external costs.	Less developed network in rural areas, well developed system of mass transit in large European cities but little optimization of time tables across transport modalities
Ownership	Private ownership and use	Public and private ownership and collective use
Costs	Expensive plus underestimation of the total cost of car ownership and use	User costs per ride are visible
Transport mode performance	Flexible (possible to combine different tasks: go to work, bring children to school, do shopping); instant or 'on call' availability	Fixed routing system and time schedule (no door-to-door service); system uncertainty of function
Cultural meaning	Part of high mobility life style; symbol of freedom; type of car signals status and expresses personal identity	Lack of privacy; seen as a functional means of transport with a bad public image
Problems	Gives rise to problems of road congestion, pollution, noise and accidents	Same as cars but to a lesser degree

maintenance (who does this, using what means), supportive infrastructures and policies, ownership structures, costs, image and performance – and the problems to which the regimes give rise (of congestion, pollution, access etc.).

Each regime has a set of rules that structures the behaviour of actors. The rules consist of product and production standards that the actors adhere to, and self-assumed roles as to what their business is. The roles and basic assumptions are a core element of a regime. The fact that car manufacturers see themselves as producers of vehicles (as opposed to mobility providers) and public transport companies see themselves as operators of a particular kind of public transport system (such as the bus system or railway system) works as an inhibiting factor for a regime-shift. These roles are sustained by the perceived need of people to own a car – and the corresponding need of having a parking place at home. The common standard for automobiles is that of an all-purpose vehicle that is over-designed in many dimensions (range, acceleration, storage space and speed) for most of the purposes for which it is used. This standard is sustained by the individual ownership and use of cars. The standard does not make sense when a neighbourhood or a car-sharing organization collectively owns cars. If cars are collectively owned, people will take a pick from the fleet of cars available, based on the individual need at that particular moment (to make a long trip, shopping trip, or pleasure ride in the countryside on a sunny day). We would have a completely different situation if the actors at the supply side were to see themselves as mobility providers offering mobility services (integrated packages of mobility) fine-tuned to individual needs. Such deep-seated changes will not come about without new technology, changes in market organization and changes in infrastructure and fiscal policies (the introduction of emission taxes and strongly differentiated vehicle duties).

Promising niches for passenger transport

Within both regimes, various technical and organizational innovations are under development to solve the problems of the respective regimes. This section will identify the most important niche developments for passenger transport. There are two leading questions for this and the subsequent section: first, do technological and market niches already exist, and second, to what extent do the alternatives have the potential to change the regime?

In the private car-based transport regime, private and public actors pursue five families of innovations to deal with issues of air pollution and /or traffic congestion. An extensive and detailed discussion of these innovations, especially their environmental score sheets, is outside the scope of this book. We can refer the interested reader to a large number of comparative studies.[9]

Environmentally improved internal combustion engine cars

Under pressure from government regulation and rising energy prices, the car industry has significantly upgraded the environmental performance and energy efficiency of its products. In Germany, for instance, the independent

Institut für Energie und Umwelt reported in 1995 that the total emissions of NO_x, CO, HC and particulates by the entire road sector had decreased since 1987, despite the growth of the vehicle fleet and traffic volume. Between 1980 and 1993, total NO_x emissions by all passenger cars dropped by 24%, CO emissions fell 38%, and HC emissions went down 30%.[10] Further reductions of emissions occurred in following years due to the continuing penetration of catalyst-equipped cars and improvements of engines and fuels, but the reductions were largely cancelled out by the growth of the vehicle fleet.

Fuel efficiency of passenger cars generally improved between the late 1970s and the late 1980s, but then worsened again more recently due to the trend towards more powerful, larger and heavier cars (prompted, in part, by the call for increased vehicle safety). For several German carmakers, Canzler and Knie (1994) have shown that new versions were in almost every case more powerful and heavier than previous versions.[11] The average fuel consumption of the new car fleet in Europe decreased by almost 15% over the period 1980 to 1995. Average consumption was 8.3 litres per 100 km in 1980 and in 1995 stood at 7.1 litres per 100 km.[12] The industry continues to work towards improvements with an emphasis on economical engines, catalysts and vehicle design (reduced drag and weight). For example, the German car industry has committed to cut the average fuel consumption of the cars sold in Germany by 25% over the 1990 to 2005 period; French manufacturers have announced they will reduce the average CO_2 emission of their cars sold in France to 150 g/km by 2005.[13] Improvements are often introduced first in high-end market segments before they trickle down to the mass market, but usually no technological niches are created (outside R&D).

Alternative fuel vehicles (AFVs)

The environmental problems of gasoline and diesel combustion and the political uncertainties of mineral oil supply have led car manufacturers and governments to look for alternative motor fuels. This resulted in various technological and market niches for alternative fuel vehicles. An outstanding example is the use of ethanol from sugar cane in passenger cars in Brazil, which had a 30% market share in the late 1980s.[14] Currently, Liquefied Petroleum Gas (LPG) and Compressed Natural Gas (CNG) are considered the most promising alternative fuels. Compared to gasoline cars, both LPG and CNG cut carbon monoxide emissions by 70–80%, hydrocarbon emissions by 70–80% and CO_2 emissions by 15–20%.[15] It is expected that these fuels will preferentially be used in high-mileage commercial fleet vehicles, where low fuel costs compensate for higher fuel storage costs – the largest obstacle to their wider adoption, along with the lack of a widespread fuel supply infrastructure.

The choice for alternative fuels generally involves no functional changes in the use of the vehicles: AFVs are specifically designed to emulate the functional characteristics of the traditional internal combustion vehicle. It is thus hard to imagine that they will contribute to a regime-shift. Both natural gas vehicles (NGVs) and LPG vehicles exist in market niches: around one million NGVs are on the roads worldwide, one quarter of these in Italy; LPG cars have a 7% market share among passenger cars in the Netherlands. Other alternative fuels that may be promoted more in the future are ethanol from biomass and methanol, which might be used for fuel cell vehicles. Hydrogen is generally considered as the ultimate clean fuel, which could be used either in fuel cells or in combustion engines. Many questions still surround the use of hydrogen as a vehicle fuel, such as how it should be produced, stored and distributed. If it is produced by electrolysis using regenerative energy sources such as hydropower or photovoltaics, hydrogen is able to mediate the greenhouse gas problem.

Electric vehicles

Electric vehicles include battery-powered vehicles, hybrid-electric vehicles and fuel cell vehicles. They have different energy sources but share the same drive system technology. Various kinds of battery-powered electric vehicles (BEVs) have been developed. Market niches have emerged for some types, such as lightweight electric vehicles in Switzerland and delivery cars in France, but most BEVs exist in technological niches (see Chapter 4). In the former case the use of BEVs has shown the promise of leading to different use patterns and integration with public transport; in the latter case the BEVs were used to replace existing fleet cars. The introduction of BEVs can thus contribute to a regime-shift, depending on the type of vehicles and introduction strategy.

Hybrid systems and fuel cells were developed to overcome the performance limitations of electrochemical batteries (for example, low energy density, power capacity and lifespan). In the case of hybrid vehicles, the electric drive system is combined with an auxiliary power unit, usually an internal combustion engine; many different configurations have been tried in recent years. Thus far, the most successful has been the system used by Toyota in its Prius hybrid already referred to in Chapter 2, which was launched in Japan in December 1997. Some 50,000 Priuses had been sold by March 2001.[16] Hybrids can easily be optimized for low energy consumption; a medium-sized car uses in the order of 3.5 litres per 100 km according to standard test cycles. There are no functional changes compared to regular cars, so that they can be introduced straight into the regime.

Fuel cell vehicles are a long-term option. Such vehicles are powered with electricity generated by the electrochemical reaction between oxygen and hydrogen. The vehicles have many advantages such as silent driving, few or no emissions (depending on the source of the hydrogen), easy maintenance, and no range restrictions. High price and safety issues currently pose a barrier to their immediate deployment. When the technology and infrastructure are fully developed, fuel cell vehicles could probably substitute for most gasoline and diesel cars in the next decades. Until then, however, they will need to be introduced in technological niches. The large car manufacturers have built several experimental fuel cell vehicles and have started field tests with these vehicles.[17]

Transport telematics systems

The application of information and telecommunication technologies to the transport domain is called transport telematics. Two categories of transport telematics can be distinguished: those that assist drivers in making decisions, leaving the control to the driver, and those which take away the control and hand it over to the technical system or its centralized management. Telematics for human augmentation is mainly inspired by traffic safety considerations, telematics for automation by traffic management (making efficient use of the road infrastructure) and environmental considerations. In particular innovations of the former kind are being introduced, generally in high-end market segments after a period of experimentation. Initial experiments with automation are also ongoing, but at present these still take place on separate test tracks and are often aimed at specific off-road applications such as automated freight transport at airports. An on-road application that is coming near to commercial deployment is so-called 'platooning', where only the first car of a train of cars is driven and the other cars follow automatically. The automobile and electronics industries are jointly developing this technology especially for trucks. Such innovations have some potential to contribute to a regime-shift, but the other telematics innovations do not, they help to optimize and sustain the existing regime.

Collective use of the private car: car-sharing

The main aim of organized car-sharing is to decouple ownership and use of cars. Users subscribe to an organization, which owns and operates a pool of cars. Access to these cars is regulated by explicit rules and trips are billed in proportion to the actual use of the system. Organized car-sharing differs from private forms (among friends or relatives) as it regulates access to the car, duties, services and rates in an explicit and professional manner. Moreover, the users get not only access to a single car, as in private car-sharing, but to the whole pool of cars owned by the organization. New reservation and accounting systems and means for access control have been invented and developed to enable these ventures to proceed. Organized car-sharing exists in market niches in several countries, in particular Switzerland, Germany, Austria and the Netherlands. In 1998, over 70,000 people were members of car-sharing organizations in these countries, and annual growth rates were relatively high in all countries (for example, in Switzerland initially over 100% and later 50–75% growth). It has the potential for a regime-shift; in particular by becoming a part of integrated mobility services, which can also include collective public transport services.

Promising niches for public transport

In the public transport regime, five niches are also considered and pursued by the companies and authorities involved.

Integration of public transport modes

Connections between different means of public transport are often cumbersome. There are few nodes for physical transfer between public transport means; passengers often lack information about the options for switching from one mode to another; and the fact that they have to buy separate tickets every time they switch modes leads to lost time and uncertainty. Increasing the attractiveness of public transport by the integration of public transport systems can be achieved by providing more physically accessible connecting nodes (stations with trains, trams, buses, bicycle hire), by providing better information about public transport services before and during a trip (travel information systems), and by using single systems of swift payment (for example, by smartcards). These three types of innovation currently exist in both technological and market niches and will generally lead to an improvement of the existing regime.[18] There is also a small potential for regime-shifts because the innovations could lead to a fusion of the regimes of individual car-based transport and public transport into a new regime.

Improved vehicle technologies

Similar to automobiles, improved internal combustion engines, alternative motor fuels and electric drive technologies are also developed for public transport vehicles. In fact, some of these technologies were first introduced in public transport, in particular buses, because replacing high-mileage vehicles offers the greatest environmental benefits especially in cities. Their large size also makes public transport buses suitable for technologies that take up a lot of space, such as compressed natural gas (CNG) (bulky gas tanks) and electric drive (bulky batteries). From a technology management perspective, public transport buses drive regular routes and return to base at the end, so that trips and maintenance can be well planned. Several market niches for alternative fuel and electric public transport buses exist.[19] They do not have the potential to contribute to a regime-shift because they substitute existing vehicles but do not change use patterns.

Self-service car rental scheme

Self-service car rental schemes involve offering cars for rent by public transport companies on a short-term basis. After use, the car is collected by the system operator and made available to the next user. Hence, use-per-car is optimized for cost effectiveness and environmental effects. Several technological niches exist in different countries where such schemes are being tested. In the case of Praxitèle and Liselec (two competing French schemes) and the Japanese 'EV+ITS' projects, the cars are small electric vehicles that are in constant communication with a central computer. Users rent the cars at special stations located, for instance, near railway stations, where maintenance and recharging also take place. Computer and communication technologies enable system operators to keep track of the location of cars and to organize distribution according to demand. The schemes have the potential to contribute to a regime-shift in urban passenger transport because they challenge ownership structures and use patterns of cars. One may look upon self-service car rental schemes as a top-down, high-tech variant of car-sharing. The Praxitèle project will be discussed in Chapter 5.

Dial-a-ride services

Dail-a-ride services provided by bus companies are another intermediate step between individual and collective transport. Passengers are picked up from home or a meeting point by minibuses and driven to their destination or to a transport node. Dial-a-ride services give further flexibility and efficiency to bus operations and prove feasible thanks to advances in information and telecommunications technology. The innovation thus has the potential to contribute to a regime-shift. Niches for such services exist in several cities in the UK, Germany and the Netherlands. The ASTI experiment in London is a good example of combining a dial-a-ride service with alternative fuel technologies – see Chapter 5.

Bicycle pools

Just as cars can be shared, so can bicycles, probably the ultimate clean transport technology. On average a journey in town by a large car uses up to eighty times more energy than a bicycle.[20] Moreover, it is virtually a zero-emissions vehicle. Offering shared bicycles on loan can be another complement to traditional public transport, especially for short distances within compact cities, on campuses, business parks, etc. Here they can contribute to a regime-shift. Such schemes have been tried in a number of European cities over the last few decades, but failed usually because of bicycles being lost or stolen. The Bikeabout experiment in Portsmouth, UK demonstrated that these problems can be overcome with new technology of chipcard operated bicycle stands, supervised by closed circuit television. Similar experiments are currently underway in Amsterdam and Copenhagen, among others.

In conclusion, some niches have the potential to contribute to a regime-shift, whereas others are most likely to reinforce the existing regime. Examples of innovations with regime-shift potential are battery-powered vehicles, telematics for traffic management, car-sharing, smartcards, individualized self-service rental systems, dial-a-ride service and bicycle pools. The last three especially have the potential for regime-shifts in urban environments for short-distance replacements. The introduction of environmentally improved gasoline and diesel vehicles (both cars and public transport vehicles), alternative fuel vehicles, hybrid and fuel cell vehicles is most likely to improve the existing regimes without change of the use context.

Future transport

The innovations discussed in the previous section are part of emerging developments put forward, in part, to improve the environmental performance of the transport regime. To help identify possible transition paths towards regime-shifts, these developments can be organized and presented in another way:

1 An environmental upgrading of internal combustion engine vehicles in terms of fuel efficiency and exhaust emissions.

2 A more efficient use and expansion of the road infrastructure through the use of transport telematics.

3 The development of market niches for alternative fuel vehicles and electric vehicles.

4 Increasing the attractiveness of public transport by integrating public transport modes by better physical connections, travel information and integrated payment.

5 The intermodal integration of public and car-based transport and the development of intermediate forms of transport, such as self-service rental schemes and dial-a-ride services.

Some of these broad lines of development are in competition with one another. For instance, internal combustion engine cars and alternative fuel cars compete directly for market share. Other paths sustain each other. Examples of mutually sustaining developments are the customization of public transport services to individual user needs, the development of transit information and booking systems, the building of park-and-ride stations and the emergence of (specialized) mobility providers.

These main development lines may add up to two different (socio-technical) scenarios, termed *regime optimization* and *regime-shift*. The starting point of both scenarios is the continued growth of mobility and equal access to necessary mobility, but not necessarily mobility by car. The first is a widely shared expectation, the other a widely shared objective and fits in the social dimension of sustainable development. In order to safeguard access to mobility, the costs of transport cannot be increased beyond reasonable limits because low-income groups would otherwise be excluded. Both scenarios are also consistent with general socio-economic trends such as income growth, increased demand for convenience, personal comfort, safety and environmental qualities.[21] We realize that other scenarios can also be constructed, including ones in which developments such as teleworking or contraction of cities play a major role. The two scenarios discussed here are not intended as predictions about the future but as constructs to organize our discussion about sustainable transport innovations. Moreover, the two scenarios could co-exist. In practice the emergence of a new regime will stimulate improvement of the existing regime, and there will be a period of co-existence of the old and new regimes.

In the first scenario of *regime optimization*, the existing cars and public transport vehicles are improved and there is only limited niche-growth, for LPG, natural gas and electric vehicles in particular. Information technology will be used to assist drivers, increasing traffic safety and utilization of the infrastructure. No co-evolutionary dynamics (including functional change in the use of the vehicles) takes place, so that no regime-shift occurs. The car-based regime continues to dominate the public transport regime. This scenario will lead to limited reductions of certain types of emissions (hydrocarbons, carbon monoxide, nitrogen oxides, and particulates) because the large reductions per vehicle are cancelled out by the growth of the vehicle fleet and vehicle mileage. The increase of mobility will also mean that other problems are not solved, such as congestion, noise and CO_2 emissions, but merely mitigated.[22]

In the second scenario existing cars are also improved, but now niche growth leads to a *regime-shift*. Such a new regime might be based on the use of a greater variety of vehicles and transport services. LPG, natural gas and electric vehicles including hybrids are introduced and are preferred over existing cars for new kinds of use (dedicated inner-city use, car-sharing, self-service rental schemes, dial-a-ride). These vehicles and services may be owned and offered by mobility leasing companies that also provide travel information and handle reservations. The companies may charge customers on a monthly basis using smartcard swipes. There are more and improved interfaces between the modes of transportation, leading to increased integration of public and private transport. A reconfiguration of transport users and alteration of travel patterns is likely to ensue when people develop new travel routines (such as increased attention to trip planning), re-assess their mobility needs and engage in car-sharing through special organizations. Some people may adopt a low-mobility lifestyle. The existing car will no longer be an all-round car but becomes a car for specific functions, especially travelling long distances. In the medium-term, fuel cell and hydrogen-burning vehicles will enter the market and compete with the (optimized) gasoline and diesel vehicles. They may also take the place of battery-powered vehicles, but for short distances BEVs may be the cheapest solution, depending on relative cost and performance of batteries and fuel cells.

The growth of the niches will lead to new economic activity by start-up vehicle manufacturers and service suppliers. But also the car industry will be prepared to co-develop the niches, because the trend in the industry is increasingly to offer cars with modular design, i.e., cars that can be equipped with varying drive systems. Since no purpose-built alternative vehicles need to be developed and produced, modular design and development allow rapid cost-reductions. The second scenario has the potential to solve most existing environmental problems. Since it will induce new production and services it will also create employment. Whether it will also solve congestion problems for cars is doubtful considering the increase of mobility, but the improvements in public transport and the integration of new transport options such as self-service rental schemes should give any traveller the option to avoid road congestion and still make door-to-door trips.

Reading these scenarios we conclude that two development lines are most likely to contribute to a regime-shift: the development of market niches for battery-powered vehicles and the development of transport means that bridge the gap between the regimes of individual car-based transport and collective public transport. Both could introduce co-evolutionary dynamics.[23] We term these two lines of development 'electrifying mobility' and 'reconfiguring mobility'.

Electrifying mobility

Electric drive technology is used in different types of vehicles. Throughout the 1990s most attention was focused on battery-powered vehicles (BEV), while hybrids and fuel cell vehicles have risen on the agenda especially in the last couple of years. The environmental aspects of these vehicles have been and are the subject of heated debate in science and policy. The benefits of current generations of these vehicles are at best limited. In the short term, therefore, electric vehicles (EVs) are not a solution to the problems of the regime; in fact, they introduce new problems, such as the disposal of batteries. But our argument is more subtle than straight substitution of technologies. New technologies will never by themselves contribute to sustainable development: this contribution depends on the quality of the introduction process and the resulting embedding of the technologies. In our view electrifying mobility is a route which has potential for sustainable development. Electric vehicles, particularly BEVs, may induce a cleaner passenger transport regime for the following reasons:

- The range restriction of battery-powered vehicles has led, in the context of the existing regime, to a deeper appreciation of the heterogeneity of the wishes and needs of travellers. It has stimulated a more accurate matching of the travellers' needs to transport technologies and services. New concepts of mobility have been developed and tested such as the stationcars concept in the US and the self-service rental schemes based on electric cars in France (Praxitèle). Gasoline cars can also serve these functions, but interestingly, the limitations of battery-powered cars have evoked these new thoughts and actions.

- The limited range of battery-powered vehicles requires users to plan their trips more carefully. Such planning reduces vehicle mileage and the number of vehicles on the road. These reductions will improve traffic flow, which has a favourable effect on both the volume of emissions and on congestion.

- The substitution of gasoline or diesel cars by zero-emission electric vehicles makes walking and cycling more attractive options for short-distance transport: pedestrians and cyclists do not have to inhale exhaust gases. These options may also come to play a more important role in new concepts of mobility, which can ultimately lead to a more effective, cleaner and quieter transport regime, especially in urban areas.

- In the realms of R&D, the limited energy capacity of batteries prompts EV developers to make the vehicle as economical as possible. The development of high-energy density batteries not only increases the driving range of EVs but also enables reduction in vehicle weight and thus energy consumption.

- The optimization of battery-powered vehicles towards energy efficiency has favourable effects on the performance of hybrids and fuel cell vehicles. Energy-efficient components and lightweight construction principles developed for battery-powered vehicles also improve the fuel efficiency of hybrids and fuel cell cars (not to mention gasoline and diesel cars). Promoting battery-powered vehicles leads to a 'favourable detour' which will eventually make way for economical fuel cell vehicles in an energy regime based on renewable sources with hydrogen as the power source. This approach also increases the chances of realizing very economic hybrids (such as 'hypercars')[24] that potentially have biofuels as an energy source.

• Discussions about battery-powered vehicles as clean vehicles have also reinforced discussions about the need to limit the emissions of electricity generation.

• EV development largely takes place outside the (more conservative) car industry. This leads to the development of a wider range of new ideas and concepts in relation to traffic and transport. The most innovative electric vehicles were developed outside the automobile industry.

In Chapter 4, we will discuss four experiments with battery-powered vehicles which, to varying degrees, have exploited the potential to stimulate a regime-shift.

Reconfiguring mobility

The prevailing mobility pattern is divided between public and private transport. Users have developed strong routines and their perceptions of performance, cost and comfort are skewed in favour of the private means of transport. Consequently, the need to own cars and the way they are used are increasingly rationalized by the prevailing mobility patterns. Public transport today either covers certain specific travel needs (like short-distance intra-urban travel or inter-city transportation) or its attractiveness is restricted to certain segments of the population that cannot afford to own or are not able to drive a car (children, elderly, disabled and poor people). The further differentiation of lifestyles and of the spatial separation of home and work place favour private transport, rather than a revival of public means of transport.

Therefore innovation routes leading to a more sustainable transport regime should include options and technologies, which do not promote additional car traffic and are able to compensate for the perceived shortcomings of public transport. This means we must look for mobility options that bridge the divide between public and private transport.

Such options can benefit from the development of information technologies. They enable the creation of new mobility forms, in particular 'integrated mobility services'. By this, we mean that there will be a departure from the one-vehicle-for-all-purposes paradigm and that specific transport means will be used for those trips for which they are best suited. Thus both public and private transport capacity may be used in a much more efficient way, while simultaneously reducing environmental impacts. Three approaches can be distinguished:

•   Initiatives to make public transport more flexible and more directly tied to the transport needs of the consumer. This includes innovations such as 'dial-a-bus' or collective taxi experiments (for example, the Dutch *trein-taxi*), which have no fixed time schedules or fixed number of collection/drop-off points. Information technology is used for route planning and vehicle tracking. The user has to learn to use this kind of system, make safe time estimates for effective use, calculate effective costs of this mobility form, and be persuaded that it represents an effective and attractive alternative.

•   Another approach is the collective use of private means of transport. Examples are car-sharing or bicycle-sharing in a community of several households in a neighbourhood, ride sharing (i.e., car-pooling, *Mitfahrgelegenheit* in Germany) or voluntary schemes for transporting disabled or elderly people. These options could be enhanced by improved information systems for interacting with the vehicles, for tracking their actual location, and for intervening in cases of malfunctions or accidents. These options rely on the adaptive capacity of users to accept that their private space becomes semi-public (as in the case of some forms of car-sharing and car-pooling) and to learn to access different kinds of vehicles for different uses. For instance, travellers have to be convinced of the reliability, timely access and appropriate transaction costs in the case of car-sharing or long-distance trip sharing.

•   A third approach to bridging the gap is actively to pressure people out of their cars and on to public transport. Here a number of inhibiting regulations can be envisioned that discourage car access to specific areas: an inner-city ban on commuters, odd-even plate access days and access only for specific vehicles like EVs, small city cars or cars with a threshold number of passengers. These schemes must be complemented by park-and-ride schemes for those drivers who are not allowed access. Information technology is likely to increase the smoothness and attractiveness of these systems. Information technology will also improve the monitoring and prosecution of offenders in case of access restrictions. However, given the stubborn adherence of many travellers to the private car, these initiatives are likely to lack widespread public acceptance and, for many, incentives to circumvent the regulations will be quite high.[25]

In Chapter 5 we will discuss four emergent technological niches (experiments) and market niches, which represent these three approaches to bridge the public-private transport gap. As for the four electric vehicles experiments in Chapter 4, we will evaluate whether these niches were successful and in what sense. We will also assess the potential of the niche development patterns for regime-shift, and argue that Strategic Niche Management is a suitable approach to exploit the potential for niche developments aimed at regime-shifts. How an innovative idea is taken up and developed and introduced experimentally is one of the factors that determines whether this potential materializes. Strategic Niche Management can orientate the niche development process to achieving a regime-shift.

Notes

1. This regime-analysis is based on Kemp, R. and Simon, B. (2001) Electric vehicles. A sociotechnical scenario study, in Verhoef, Erik and Feitelson, Eran (eds.) *Transport and the Environment: In Search of Sustainable Solutions*. Cheltenham: Edward Elgar.

2 The pre-eminence of car-based travel patterns is based partly on the assumption that ownership and use of a means of transport have to be tightly coupled in order to deliver an adequate level of utility to the user. Once the huge upfront investment in buying a car has been made, its availability in front of the house creates an incentive to use it as often as possible. Costs calculated on a per-kilometre basis are likely to decrease with increasing yearly distance driven by the car.

3 Nieuwenhuis, Paul and Wells, Peter (1997) *The Death of Motoring. Car Making and Automobility in the 21st Century*. Chicester: Wiley, see the mass production of sheet-metal bodies as the core of the automobile regime. They argue that the car manufacturers are trapped in a vicious cycle of having to make huge investments in sheet metal body production, which, as the profit margins in the very competitive car industry are small, can only be earned back if the vehicles are sold in mass. Consequently, the industry prefers to employ conservative designs and technologies. They advise car manufacturers to change to smaller batch production methods to increase profitability, and use the leeway created to build 'environmentally optimized vehicles'.

4 Vester, Frederic (1990) *Ausfahrt Zukunft – Strategien für den Verkehr von Morgen.* München: Heyne Verlag, argues that these dimensions are derived from the central requirement in car design, i.e. speed. The high-speed requirement leads to large and heavy vehicles (for safety reasons), consequently with high fuel consumption, high emission levels, unsafe traffic (long braking distance), inefficient road use etc.

5 See Achterhuis, Hans and Elzen, Boelie (eds.) (1998) *Cultuur en Mobiliteit*. Den Haag: Rathenau Instituut, especially the contribution by Tieleman, Henk, Modern Mobiel, pp. 85–102, which deals with the broad cultural meanings of mobility.

6 Ministry of Transport (1993) *Annual Report on the Transport Economy; Summary (Fiscal 1992)*. Tokyo: Ministry of Transport, p. 11.

7 According to Maddison *et al.* (1996), the total costs of road transport in the UK in 1993 were between £ 45.9 and 52.9 billion with a taxation return of £ 16.4 billion (quoted in Whitelegg, John (1997) Finding an Exit from the Mobility Maze: Non-conventional Approaches to Mobility in Urban Areas. IPTS Report No. 11, pp. 23–28).

8 There is an interesting dichotomy in the perception of public transport between men and women. Whereas men scorn the lack of privacy of public transport, women tend to enjoy the company of other travellers.

9 See, among others, Whitelegg (1993), *op. cit.*; Berger, R. and Servatius, H.G. (1994) *Die Zukunft des Autos hat erst begonnen. Ökologisches Umsteuern als Chance*. München: Piper;

Riley, R. (1994) *Alternative Cars in the 21st Century: A New Personal Transportation Paradigm*. Warrendale: Society of Automotive Engineers; MacKenzie, J.J. (1994) *The Keys to the Car: Electric and Hydrogen Vehicles for the 21st Century*. Baltimore, MD: World Resources Institute; Sperling, D. (1995) *Future Drive: Electric Vehicles and Sustainable Transportation*. Washington DC: Island Press; Nieuwenhuis and Wells, *op. cit*. Important sources are also the reports on low-emission vehicles and clean fuels for the California Air Resources Board. A Dutch comparative review of new propulsion systems is the report by Smokers, R. *et al.* (1997) *Verkeer en vervoer in de 21ᵉ eeuw, Deelproject 2: Nieuwe aandrijfconcepten*. TNO-Automotive, December.

10 The study was cited in Zimmermeyer, Günter (1995) Das Elektrofahrzeug – seine Perspektive aus der Sicht der Autoindustrie, in DGES (1995) Fachtagung 'Elektrofahrzeuge im Aufschwung – Zukunftsvisionen und Realität. Berlin, April.

11 Canzler, W. and Knie, A. (1994) *Das Ende des Automobiles, Fakten und Trends zum Umbau der Autogeschellschaft*. Heidelberg: Verlag C.F. Müller, pp. 56–64.

12 ECMT (1997) CO_2 *Emissions from Transport*. Paris: OECD, p. 182.

13 ECMT, *op. cit.*, p. 179.

14 Morales, R.M., Storper, M. *et al.* (1991) *Prospects for Alternative Fuel Vehicle Use and Production in Southern California: Environmental Quality and Economic Development*. Los Angeles: University of California, Lewis Center for Regional Policy Studies, pp. 11–12.

15 Elzen, B., Hoogma, R. and Schot, J. (1996) *Mobiliteit met Toekomst. Naar een Vraaggericht Technologiebeleid*. Report for the Dutch Ministry of Traffic and Transport. The Hague: Ministerie van Verkeer en Waterstaat, p. 29.

16 Toyota Sells 50,000 'Prius' Hybrid EVs, 8,000 in U.S. *Calstart News Notes*, 26 March, 2001.

17 Several informative websites on this subject have sprung up in recent years, such as www.calstart.com and www.hyweb.de.

18 See Elzen, Hoogma and Schot, *op. cit.*

19 Examples are CNG buses in France, Germany and other countries; bioethanol buses in Sweden; and LPG buses in the Netherlands. The size of these niches is in the order of several hundred vehicles.

20 Potter, Steve (1997) *Vital Travel Statistics*. London: Landor.

21 Similar scenarios can be found in Elzen, Hoogma and Schot, *op. cit.*

22 For congestion there is no long-term technical solution, building extra roads is expensive and increasingly difficult in built environments. The solution to congestion lies in managing traffic flows and encouraging people to change their mode of transport or travel less.

23 For this conclusion see also Elzen, Hoogma and Schot, *op. cit.* They used the notion of key technologies and identified the second development line with mobility information technologies.

24 A *hypercar* is a hybrid concept in which economy of weight is the most important design criterion. Using composite materials, a four-person car might weigh as little as 400 kg. Aerodynamic design, low-friction tyres, and recycling of brake energy would make such a vehicle very efficient. On the basis of demonstrated components, the fuel consumption of such a vehicle could be less than 1 litre/100 km. The hypercar concept was developed outside the car industry and has not been demonstrated, although some prototypes by small Swiss companies come close to it. See Lovins, A.B., Barnett, J.W. and Lovins, L.H. (1993) Supercars. The Coming Light-Vehicle Revolution. Paper presented to the Summer Study of the European Council for an Energy-Efficient Economy, Rungstedgård, Denmark.

25 The massive infringements of the access rules in the case of limited access zones in Italian city centres illustrate this quite impressively. In Bologna the number of vehicles that had unauthorized access before an automated control system was implemented was estimated at 30–40%. Hoogma, R. (1998) Introduction of Automated Zone Access Control in Bologna, Italy. A Case Study for the Project 'Strategic Niche Management as a Tool for Transition to a Sustainable Transportation System'. Enschede: University of Twente.

Chapter 4

Experiments in electrifying mobility

In the early 1990s, experiments with electric vehicles (EVs) became popular in several European countries and in the United States. They all focused on battery EVs but also represented fundamentally different approaches to the technology itself as well as to experimentation with, learning from, and institutional embedding of new transportation technologies. Because these programmes aimed to reduce the environmental impacts of transportation, they represent excellent opportunities for exploring the articulation of Strategic Niche Management policies.

In the first part of this Chapter, we provide the background history of EVs. In the second part, we present and analyse four experiments in detail: in Rügen (Germany), La Rochelle (France), Mendrisio (Switzerland), and Oslo/San Francisco. In the final section, we analyse the successes and failures of these experiments in terms of their contribution to niche development, using the Strategic Niche Management perspective. We conclude with a discussion of the added value of Strategic Niche Management.[1]

The early years: competition with gasoline cars[2]

Electric vehicles have been around for a long time. They first appeared on the scene in the late nineteenth century. Steam and petrol engine vehicles initially figured as competing alternatives. Many experiments were undertaken. In 1890, the most important means of land transportation were the steam train and the urban horse-drawn tramway, which had almost replaced the horse-drawn omnibus. Because horses caused many nuisances and were expensive to keep, there was a latent demand for an alternative.[3] Moreover, the demand for transportation was as a result of increasing urbanization. Alternatives to horse-drawn public transport became available with the introduction of electric trolley buses and bicycles. Bicycles were introduced in several niches. Racing was important, and speed soon emerged as an important design aspect. Touring became a second niche. Although European city administrators were reluctant to introduce electric trolleys because the overhead lines disfigured the landscape, their counterparts in the United States adopted them quickly, with the active support of electricity producers. The electrification of tramways was fuelled by the *'universal enthusiasm for electrical technology'* in the United States in the late nineteenth century.[4] In Europe, a brief and ultimately unsuccessful period of experimentation with battery tramways ensued, but the fumes from the lead-acid batteries were too unpleasant for passengers.

These new means of transportation were implicated in a transport revolution out of which grew a whole new set of user preferences, institutions, and government regulation. A new culture of mobility emerged that included an appreciation of speed, touring, and individual freedom. Electric road vehicles were developed as derivations of the electric trolley. They used some of the electric trolley technology, notably batteries (although electric trolleys got electricity from powerhouses, i.e., stationary batteries), battery chargers, DC electric motors, and mechanical controllers. Because these vehicles used the roads and not rails, their drive systems had to be redesigned to meet the requirements of driving on roads with untrained drivers. The vehicle design was based on the horse carriage, but later converged with designs based on bicycles because many early manufacturers of both gasoline and electric cars had initially built bicycles.

In those days, EVs were considered clean, quiet, reliable, and easy to handle. Performance characteristics such as 'low' maximum speed, the 'heavy' weight of batteries, and the vehicles' 'limited' range, which are usually emphasized in today's discussions, initially were not of central importance. The 'use environment' of the technology was very different. The prevailing infrastructure of roads was not optimized for car use; users' primary needs were local mobility; and opportunities for electric touring were limited at best. The major advantages of electric vehicles included their reliable performance, ease of starting, and ease of maintenance, except for the battery. Gasoline cars, on the other hand, required considerable force to 'crank' the motor and were unreliable in use, thus requiring specialized mechanics and drivers who could maintain the vehicles on the road. These aspects, when compared to those of electric vehicles, made gasoline cars quite unattractive for some users. The range limitations of electric vehicles were recognized early on, so that their application was limited to cities. Initially there were very high expectations about the potential for EVs in city use. Fogelberg quotes a contemporary observer:

> Anyone who is familiar with the condition of the art and with the character of the product of the various types of motor vehicles [i.e. steam, petrol, electric], cannot doubt the wide field [i.e. use in cities] that the electric vehicle will cover. [5]

Gasoline vehicles become dominant

Historians are still debating the relative importance of the various reasons why gasoline vehicles became more widely used than electric vehicles. In the context of this book, a full explanation is not necessary. Following our theoretical model, this would require an analysis of niche developments, a regime analysis, and a study of developments on the socio-technical landscape such as the emancipation of women, the isolation of American farmers, and the process of urbanization. However, we would like to indicate some niche developments to provide context for recent attempts to electrify mobility.

Steam vehicles were introduced on the streets before 1890, but suffered from association with the common and dirty train, an old-fashioned technology. They also met strong resistance because of the fear of boiler explosions.[6] Electric vehicles were ideally suited for the urban transport niche where range limitations posed no problem. EV manufacturers anticipated growing demand for luxurious short-range urban vehicles to replace horse-drawn carriages. Therefore they modelled their products after the already-familiar horse-based technology, and initial acceptance of electric vehicles was better than that of the other alternatives because of this resemblance. The initial purchase cost of electric vehicles was much higher than that of gasoline vehicles because of the expensive batteries, but electric vehicle producers did not perceive this as a problem because electric cars were usually more exclusive and targeted at a different market.

One of the first niches for the development of the gasoline car was leisure touring in the countryside, which built on a tradition introduced by the bicycle. Gasoline cars evolved towards speed and endurance, which gave them an image of youthfulness, adventure, and novelty. The many speed and endurance races (a second niche), very popular in the late nineteenth and early twentieth centuries, further established the gasoline car as the embodiment of the characteristics that a vehicle should have. Although electric cars performed better in short-distance, high-speed races, these were much less popular than the endurance races at which the gasoline vehicles excelled.[7] The latter races also provided a stabilized design of 'the passenger car', which was based on the 1901 Mercedes model. The emergence of a basic design for gasoline cars and the lack of one for electric cars is yet another factor that explains the eventual dominance of the gasoline car. *De facto*, the Mercedes cars set the standard for engine design and vehicle technology on which the development and manufacturing principles for cars became based.

Concurrently, the context in which cars were used changed as, increasingly, they were seen as a means for travelling long distances and achieving potentially high speeds.[8] Electric vehicles could not keep up with these developments. They soon lost their initial advantages as hundreds of variants were built at high cost, while the new use context did not fit their performance characteristics. Also, electric touring vehicles suffered from the lack of reliable access to charging facilities. Especially prior to the introduction of current inverters in 1905, electric vehicles could be used only in areas with direct current and were excluded from areas served by alternating current.[9]

After gasoline had established itself in several niches, it started to invade the niches occupied by the electric car. For example, David Kirsch shows that in a number of US cities, electric vehicles helped to build acceptance for motorized horse-drawn carriages, which were initially banned from using boulevards within urban parks. After the EV won over the sceptics, steam and gasoline vehicles were also permitted to share the roads with horses.[10] Later, after gasoline cars incorporated electric starters and with the growth of a dense network of refuelling stations, these cars became ever more attractive to less adventurous users.

By 1907-1911, it became clear that the gasoline car had outcompeted electric vehicles for most applications. After World War I, a bandwagon of gasoline car diffusion set in, with intensive investment programmes in road development linked with the construction of mass-production plants for cars and state housing programmes. Within a matter of decades, these processes changed the face of the urban landscape more profoundly than any developments for centuries before. The gasoline car became the paradigmatic product of an entire era.[11] Nevertheless, the electric vehicle would thrive for some years to come in a number of niches.

Remaining niches for electric vehicles

From the start, electric vehicles were developed for commercial passenger service rather than for private touring. However, the leading American electric cab venture collapsed when the operating company became overextended and could not deliver high-quality service in all the cities it had promised to serve. Furthermore, the company was struck by a corruption scandal that led investors to withdraw from the project. These events also had a negative influence on European developments, where expectations were also initially high; many European cities favoured electric taxis over electric trolleys because the former did not require unsightly overhead lines. In spite of these setbacks, however, electric cabs were used profitably in Amsterdam until 1920.[12]

Commercial vehicles were another niche application for electric vehicles. By 1913, 40% of trucks in New York City were electric, most of them having replaced horse-drawn delivery wagons. One reason for the popularity of electric vehicles with businesses was that they were cheaper to use than horses and gasoline cars. Around 1910, the price of electricity began dropping while gasoline started becoming more expensive. A 1913 MIT report comparing the relative costs of commercial vehicles propelled by electricity, gasoline, and horses showed that the horse was somewhat cheaper on a cost-per-day basis, but the cost per mile driven or cost per delivery made was 75–85% lower for electric vehicles than for gasoline cars and horse-drawn vehicles.[13]

Another advantage of electric vehicles was that they were more efficient in tasks that required frequent stopping and starting. Such work was very demanding for horses, but electric vehicles were quicker in stop-and-start traffic than gasoline cars, due mainly to the higher torque of electric motors and the fact that the electric vehicles did not use multiple gears. In addition, the absence of emissions was used as an argument in favour of using electric vehicles for food delivery. By 1914, the American electric vehicle fleet included nearly 40,000 vehicles, compared to only a few thousand in Europe.[14]

Nevertheless, the market niche for electric commercial vehicles shrank as cities grew and distances to the suburbs increased. Despite technical improvements to new generations of electric trucks (e.g., up to 50% higher speeds, increased driving range, lower purchase prices), fleet operators increasingly opted for gasoline trucks. According to Mom, electric traction disappeared from public roads and went 'underground' as electric vehicles were used as industrial trucks inside and around factories and in harbours. The German postal service continued using EVs on roads (before World War II, the majority of its fleet vehicles were electric), and the UK developed a fleet of several thousand electric utility vehicles that included post vans, trolley buses, garbage-collection trucks, and delivery trucks.[15] Such applications proved to be excellent niches for electric vehicles because high driving speeds were not necessary and their silence was a benefit. In several European countries, electric vehicles regained some importance during World War II when fuel was scarce, but in the United States, 'the electric automobile industry, having been in slumber for two decades, wiggled a little in 1955, then again slept soundly'.[16] The environmental movement of the 1960s and the OPEC oil crisis of 1973 would awaken it again.

In addition to industrial applications such as forklift trucks, electric vehicles also continued to be used in specialized market niches such as milk carts in the UK and indoor-transport vehicles at airports and train stations. More recently, the electric vehicle has been adopted as a quiet and non-polluting means of transport in holiday resorts such as car-free Swiss mountain villages and in US retirement communities.

Resurgence of interest in electric vehicles

As mentioned above, interest in stand-alone electric-powered vehicles resurfaced in the late 1960s and early 1970s. Several research institutes and firms in Europe and the US started projects on battery electric vehicles and conducted a number of field tests. The results were quite sobering, for the vehicles had poor technological performance and were far too expensive compared to the standards set by the gasoline car. By the mid 1980s, however, alternative modes of transport had gained widespread attention due to the growing environmental movement, the worsening air pollution problems in urban areas, and, later still, the issue of global climate change, which attracted increasing interest in public debates. Also, in some countries, public distrust of nuclear power motivated active searches for energy-supply systems based on renewable energy sources and strongly influenced the political environment.

This increased political attention resulted in calls for the development of more environmentally benign cars, among them electric vehicles. We have identified three different approaches that have been used to try to bring electric vehicles onto the market:

- A top-down approach, in which policy-makers tried to force industry to develop and introduce electric vehicles;
- An endogenous approach within the automobile industry; and
- A bottom-up approach initiated by concerned citizens.

The California Air Resource Board (CARB) pursued the most aggressive top-down approach through its ZEV (Zero Emission Vehicle) mandate of 1990.[17] The mandate required that by 1998, zero-emission vehicles would have to make up 2% of all vehicles sold in the state by large-volume car producers. From 2003 onward, even medium-sized car producers were to be held to the same requirements. Failure to comply would result in a $5,000 fine for every electric vehicle 'not sold', although many observers doubted whether that penalty would actually be enforced. CARB expected that the mandate would put 400,000 ZEVs on the roads by 2003, and, according to interpretations of the language of the mandate, only electric vehicles would count as ZEVs.

Much of California's interest in electric vehicles stems from the state's failure to meet federal and state air-quality standards in many (especially urban) areas. In southern California, for example, specific climatic and geological circumstances combined with high population density have made the atmosphere in the region one of the most polluted in the world. As a result the public is willing to accept stringent pollution-control policies. In addition, California has an established tradition of enacting legislation to force industry to curtail automobile emissions.[18] CARB hopes to reduce local emissions substantially by forcing electric vehicles onto the market. Even if emissions from power plants are taken into account, the well-to-wheel emissions of electric cars in California are still substantially lower than those resulting from gasoline cars. Californian power plants have been under stringent emissions control for quite some time. These existing regulations have favoured the use of natural gas in a large fraction of power plants, and the trend is away from coal and oil.

Economic interests are also important for understanding the CARB mandate, for California lacks a major car industry that could portray the measures as unattainable or harmful to the local economy. On the contrary, the state sees the production of electric vehicles and their components as an important future industrial alternative for the retooling military industry and for innovative new local firms. In addition, such new industries would profit from the enormous car market in the region, which absorbs about 4% of world car production and up to 10% of the production of specific manufacturers.[19]

In essence, with the ZEV mandate, the California State government promised a market for electric vehicles in the near future. This led to a bandwagon of initiatives by various actors as electric utilities, governments, and corporations, both inside and outside the United States, committed funds to support development of vehicles and components, mainly to the benefit of small R&D companies. Many actors also added experimental and pre-production electric vehicles to their fleets and invested in recharging facilities to stimulate the development of a market. Several other states have followed California's example and adopted the mandate too. And at the federal level, the Department of Energy instigated the well-funded US Advanced Battery Consortium with participation of the major car manufacturers, the battery industry and electric utilities, who targeted the development of a high-performance battery.

In Europe, too, electric utilities and cities have adopted electric vehicles in their fleets. The French state electric utility EDF operates close to 2,000 EVs. Some seventy cities formed the association Citelec in 1980 with the aim of actively promoting the use of electric vehicles. Various EC-funded demonstration programmes have helped increase the use of electric vehicles. One such programme is ZEUS, in which eight cities jointly procured several hundreds of electric vehicles, as well as vehicles running on biogas and natural gas.[20]

The *endogenous approach of the automobile industry* was a reaction to public criticisms of the automobile. The industry considered electric vehicle technology potentially interesting, although it proved to be fundamentally different from the prevailing technological regime with regard to both technology and manufacture. After CARB announced the ZEV mandate, the automobile industry began developing electric vehicle prototypes of ever-increasing technical sophistication and improved performance characteristics. CARB considered the involvement of the car industry crucial if mass-produced and thus affordable vehicles of high quality were to enter the market.[21] In reality, industry-produced EVs were often expensive because the manufacturers oriented their activities to replicating the performance characteristics of existing gasoline cars. The best-performing but also most expensive examples were purpose-built cars such as General Motors' Impact/EV1 and Honda's EV-Plus.

Ironically, the Impact was actually the vehicle that triggered the ZEV mandate. When General Motors presented the prototype at the January 1990 Los Angeles Auto Show, it made electric vehicle technology seem real and encouraged CARB to include electric vehicles in the mandate it adopted in September of that year. Although this result was foreseen by some GM officials, it pleased neither the company nor its Detroit competitors and precipitated intense controversy about the mandate. However, even as the industry attempted to mobilize public opinion against electric vehicles, there was a feeling of inevitability regarding the introduction of electric vehicles. When CARB relaxed the mandate in 1996, the carmakers announced that they would put their prototypes on the market, albeit in small numbers until better batteries were developed.[22]

Most car manufacturers favoured electrifying existing models as an initial, low-cost strategy, but even these vehicles were still more expensive than their conventional counterparts because of both the price of batteries and the small production volumes. Some manufacturers opted for advanced batteries, which offered the highest performance in terms of energy content. Although the German companies used these batteries only for testing purposes, some Japanese companies also offered advanced battery equipped vehicles for sale. Electric vehicles from Toyota, Honda and Nissan were generally reckoned among the best available, but they were also among the most expensive, and only a few hundred of each of these cars were sold.[23]

Whether purpose-built or converted from existing designs, few electric vehicles were introduced because the car industry had no faith in the existence of demand. The car manufacturer that showed itself to be the most serious about commercializing battery vehicles was PSA (Peugeot-Citroën S.A.), which sold several thousand electric vans and small passenger cars.[24] Fiat also made an early decision to market conversion-type electric vehicles. Because of the vehicles' high price and experimental nature, but also as part of a deliberate marketing strategy, these manufacturers sold most of their EVs to fleet operators.

The *bottom-up approach* has been followed by a wide variety of outsiders. Almost every major industrialized country has seen initiatives for reinventing the car. However, the most promising experiences were in countries that lacked an indigenous automobile industry, namely Denmark, Switzerland, and Norway. Denmark actually produced most of the electric vehicles currently in use in Europe. The most popular model was the City El, a small, three-wheeled one-seater. With a top speed of 50 km/h and a driving range of 50 km, its performance is modest but adequate for urban trips, and, at 12,000 DM, it is affordable compared to other electric vehicles. Between 1987 and 1995, 4,500 City Els were built and sold; by then, they made up about half of the electric vehicles on the roads in Europe.[25]

The largest per capita market for electric vehicles emerged in Switzerland, where a number of small and medium-sized enterprises, university institutes, interested individuals and spin-off firms founded by students from technical universities built up an innovation network to develop and diffuse lightweight electric vehicles (LEVs). These endeavours found strong support from the expanding anti-nuclear-power movement of that time. By the mid 1980s, two initiatives had taken shape. One was the consolidation and professionalization of the open network of pioneers (i.e., amateurs and small start-up firms) referred to above. The other was the start of the so-called 'Swatch-Mobil' project.

Originally, the pioneers' products were known as 'solar vehicles' and consisted of bicycle components, wood, electric drives borrowed from other products, and added-on solar panels. Eventually, the solar panels disappeared, the vehicles improved in quality and performance, and some of the most pressing technological problems surrounding the application of lightweight construction principles and materials to vehicles were solved.[26] The pioneers' beliefs took shape in the 1970s with the OPEC oil crisis and the publication of *The Limits to Growth*, the report to the Club of Rome that questioned the unlimited availability of oil resources and warned of the high vulnerability of modern mobility.

The Swatch-Mobil was initiated by the Swiss watch industry through the SMH (Société Suisse de Microélectronique et d'Horlogerie SA), its most important holding. SMH saw a parallel between their own success with the radically new concept of the Swatch wristwatch and the radical innovations in automobile technology emerging out of the network of individual developers. Thus, SMH became interested in LEVs and launched their own development plans.[27] The Swatch-Mobil, a small electric vehicle that was to be sold at the relatively low price of about 10,000 Swiss Francs, was introduced in the early 1990s. Later, SMH brought its know-how into a joint venture with Mercedes-Benz (now DaimlerChrysler), which brought the 'Smart' car to the market. While it is certainly an innovative product in terms of design, advertising, production system, marketing-by-franchising strategy, and pool leasing service, it is technically conservative, as it is equipped with a conventional combustion engine. For this reason, in 1999 SMH sold its shares in the company that produced the Smart, claiming that the joint venture had failed to achieve their original goals of developing a radical ecological innovation.

In the beginning, all these developments happened independently, but soon it became necessary to share experiences, concepts, and plans actually to test the new vehicles. One of the most important events in the formation of the social network around LEVs was the 'Tour de Sol'. This race, held from 1985 to 1992, consisted of a weeklong competition for prototypes and commercially available lightweight electric vehicles. The route crossed Switzerland's entire national territory and included some steep mountain roads. A yearly national conference and a solar technology fair were held concurrently with the Tour de Sol to exchange information between developers and to inform international experts of developments in

Switzerland. Also important was a programme of crash tests during which problems of passenger safety were identified and resolved.

The enthusiasm generated by public events like the Tour de Sol stimulated the market for LEVs. Several firms began importing small cars such as 'voiturettes' and equipping them with electric engines. Voiturettes are small, light, two-seat sub-compact cars that in France can be driven without a driving licence by anyone of 14 or older. With a maximum speed of 45 km/h, they are easily distinguishable from 'normal' cars by their tiny size and particular shape, and they also enjoy special tax benefits. These 'conversions', together with the three-wheeled Danish City-El, form the great majority of LEVs currently being driven on Swiss roads.

As the early market developed, so too did a network of selling and servicing firms to support the daily use of these vehicles. This network benefited from the technical knowledge of the converters and the prototype developers, who in turn were interested in the everyday reliability of LEV components and vehicle concepts. The experiences of early LEV-users thus became important inputs for the development of further prototypes. Most prototypes, however, were never fully commercialized.

Slowly, a 'proto-market' for LEVs established itself. More than 2,000 LEVs had been sold in Switzerland by 1995. Before the large-scale Mendrisio experiment started, Switzerland already represented about 20% of the world market for electric vehicles. In the early 1990s, the market experienced rapid growth. In 1992, however, sales figures began to fall and by 1995 had stabilized at between 100 and 200 vehicles per year (see Figure 4.1). This

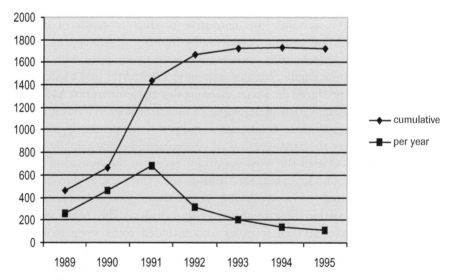

Figure 4.1 The LEV market until 1995, the year the large-scale demonstration programme began

slowdown had several causes. The Swiss economy entered a long period of stagnation, and sales of many innovative products, including new automobiles, experienced a substantial decline.[28] In addition, the quality of the early LEVs was poor. The vehicles did not fulfil minimum standards of reliability and were costly both to buy and to maintain. In addition, potential consumers held back from buying new LEVs after SMH announced plans for a low-cost, high-performance LEV. Finally, by the early 1990s, the environmental movement lost some of its earlier momentum.

All three of these approaches contributed to the increased interest in battery electric vehicles (BEVs) from the early 1980s to the mid 1990s. For some time, battery electric vehicles were considered the most promising path towards a more sustainable transport future. The bottom-up developments in Switzerland, the prototypes of the automobile industry, and the ZEV mandate in California created expectations about the marketability of electric vehicles among a variety of actors and led them also to become involved in initiatives to develop and market electric cars. Thus, a number of experiments were undertaken in different cities and regions. To these experiments we now turn.

Strategies for Experiments

During the past decade, representatives of all three of these approaches – policy-makers, the automobile industry, and outsiders – initiated a wide variety of experiments with battery electric vehicles. This variety was amplified by the fact that the experiments tended to be carried out in widely differing geographic and historical contexts.

Two fundamentally different technological strategies have characterized recent experiments with BEVs: conversion design and purpose-built design.[29]

Conversion design replaces conventional internal combustion propulsion systems with an electric motor. The resulting product may best be described as an 'electric car'. Because this approach essentially involves exchanging parts while keeping the basic vehicle concepts intact, this strategy has definite advantages in that it is compatible with the dominant transport regime. It builds on the competencies, infrastructure, and routines of existing networks of manufacturers, users, and regulators. And it was, therefore, the strategy favoured by those actors most entrenched in the dominant regime, i.e., the automobile industry. From the standpoint of sustainable transport, however, conversion design has some definite disadvantages. For one thing, the car's heavy weight requires large battery packs to achieve satisfactory driving range. But the batteries add to the weight even more, resulting in higher costs and lower energy efficiency.

Purpose-built cars, on the other hand, get the best from their electric drive by reducing vehicle weight and size and optimally integrating the power source into the vehicle structure. To compensate for the BEV's technical shortcomings, reducing energy consumption became an important goal, and improvements in energy efficiency of a factor 3 to 4 have been achieved compared to conventional vehicles. Examples of the purpose-design strategy are the Swiss developers' lightweight electric vehicles, small three- or four-wheeled vehicles that can carry one to four passengers. Their construction relies on the use of lightweight materials (for example, fibreglass bodies) and lightweight construction (for example, aluminium frames). For these reasons, the LEVs are extremely energy-efficient compared to a conventional car (approx. 100-200 Wh/km, roughly equivalent to one to two litres of gasoline per 100 km).[30]

Various models of LEVs differ in technical performance. The most recent prototypes have a driving range of about 200 km per battery charge, a maximum speed of 120 km/h, and a recharging time of only a few hours. Older models have a range of only 30 km per battery charge, a maximum speed of 50 km/h, and need about eight hours to recharge the battery.

Nevertheless, small and lightweight battery-powered vehicles have their drawbacks. They are not easily integrated into the manufacturing and service infrastructure of the dominant gasoline car regime, so owners often encounter resistance from manufacturers, dealers, regulators, and users who denounce such vehicles as 'not real cars'. Nor do these vehicles conform to the selection criteria set by the regime. In particular, critics often raise safety issues and claim that the occupants of small and light vehicles have little chance of surviving a crash with a 'regular' car. To counter these criticisms, lightweight vehicle developers have implemented new and effective safety mechanisms that have been shown to protect passengers in collisions. The Swiss developers have especially distinguished themselves in this respect.[31]

These two technological development strategies for electric vehicles parallel another two-fold dimension related to the different conceptions of the users of an automobile, namely who will be the users and what are their needs, and how the vehicles will be perceived and used. Thus, the conversion strategy first takes the user characteristics of a conventional gasoline car as given and tries to fulfil these expectations as closely as possible, while the purpose-built strategy tries to develop a completely new product identity for the electric vehicle and projects a market where new kinds of users and applications emerge.

Combining these two dimensions, we can identify four possible experimental strategies for introducing electric vehicles (see Table 4.1).

Substitution combines conversion design strategy with relative inattention to user needs. BEV drivers are expected to have similar needs to users of gasoline cars, and the reference point is the prevailing gasoline car. This leads to an emphasis on the shortcomings of the electric vehicle, including limited driving range, long recharging times, high price, and substandard quality of finish. This approach thus emphasizes developing a new kind of battery and upgrading other vehicle components to enhance overall performance.

Leapfrogging aims at developing a vehicle with revolutionary design that can compete directly with conventional cars, thanks to some specific features not offered by conventional cars. It will be introduced initially in specific market niches, but is not restricted to them.

Market differentiation combines a converted vehicle with the development of a market niche where users will be less sensitive to the shortcomings of the electric car and might therefore appreciate some of its specific features. The emphasis is on identifying users' needs and testing user acceptance in specific market niches.

The *new mobility option* combines radical design with an emphasis on exploring user needs. Here the vehicle is only one part of a broader array of transport options that are developed concomitantly with the new vehicle. The innovation process is probably the most complex of the four options.

Table 4.1 Experimental strategies for battery electric vehicles as a contribution to sustainable transport future

		Conception of user	
		Users taken for granted	Exploration of user needs
Conception of artefact	Conversion design	**Substitution** (Rügen)	**Market differentiation** (La Rochelle)
	Purpose design	**Leapfrogging** (PIVCO)	**New mobility option** (Mendrisio)

The next section of this Chapter analyses four experiments with electric vehicles that cover the four paths mentioned above as summarized in Table 4.1.

◆ The *Rügen* project involved a field test of 60 conversion-design electric vehicles made by different German manufacturers, equipped with advanced, high-performance battery systems, and used as replacements for regular cars. Users were local companies on the German island of Rügen.

◆ The *La Rochelle* project included several experiments with conversion-design electric vehicles developed by the car manufacturer PSA and carried out in partnership with the electricity producer EDF and the municipality. The cars were marketed as second cars for urban professional and private users. The experiment on which we will focus involved 50 vehicles and was designed to test user needs.

◆ The *PIVCO* experiment involved the *City Bee/TH!NK* car, developed by the Norwegian firm PIVCO using innovative materials and design principles and intended mainly for urban use. The users of some 100 experimental cars in various experiments included fleets, car rental users in Oslo, and users of station cars in northern California.

◆ The *Mendrisio* experiment aimed at introducing 350 electric vehicles into a Swiss region. Targeted drivers were mainly second-car users, but alternative ways of using the vehicles, i.e., different use patterns, were also addressed in accompanying studies.

These four cases are each represented in a specific cell of Table 4.1. However, these locations are not fixed; each experiment entailed a multitude of parallel goals. There is no easy fit for any of the cases, as many of the experiments evolved over time and were subject to changing external conditions. Even so, different emphases can be discerned.

Each case study is presented according to the following basic format: three sections introduce each experiment and provide background followed by sections that discuss features of niche development processes inspired by the theoretical analysis presented in Chapter 2:

◆ Setting and context of the experiment: description of background, main initiators, and their motivations to carry out the experiment, along with a brief history of the niche development.

◆ Objectives and project organization: description of the actual project plan, objectives, the main actors associated with the experiment's organization, the choice of location, and the main funding sources.

◆ Project management: discussion of relevant developments in the experiment and project management, including whether the goals were formally achieved, the major difficulties encountered, how the experiment was overseen and its outcomes communicated to broader external constituencies.

◆ Contribution to learning in the niche: analyses of niche development, including focusing first on the nature of the learning process. Questions asked include: what kind of learning took place, and on which aspects? Can we see any signs of second-order learning (and co-evolutionary dynamics), for example, the emergence of new ideas about future mobility? If so, how were these outcomes captured by the project management? In this area in particular, it is important to look at how users were integrated into the project.

- Contribution to institutional embedding in the niche: institutional embedding refers first of all to the quality of network development and is targeted at answering the following questions. Who was involved, only insiders, or were outsiders also welcomed? What did these actors contribute to the process? To what extent do we have a highly aligned network? This section then discusses whether expectations became more widely shared. And finally, we look at the development of new infrastructure and production sites that may provide a set of complementary assets to aid in further development of the niche.

- Epilogue discussing future prospects.

Rügen Island: testing the latest components[32]

Introduction and background

One of the most important and visible testing and evaluation projects of electric road vehicles was launched in October 1992 on the German island of Rügen. For nearly four years, more than a hundred users tested 60 prototype electric vehicles in an experiment organized by the German automobile industry and battery manufacturers under the auspices of the Federal Ministry of Research and Technology. The vehicles included passenger cars, vans, and medium-sized buses, all based on advanced battery systems and electric motors from the participating manufacturers.

The experiment took place against the following backdrop: the Federal government viewed electric vehicles as a technology that offered few benefits for either users or society. Because coal-fired power plants produced 55% of the country's electricity, the Federal Environment Ministry believed that there was little to be gained by using electric vehicles. The benefit of reducing local emissions would not offset the costs of emission increases at the power plants, except in specific places such as health spas, hospitals, and in pedestrian zones. The Ministry also argued that batteries were too heavy, so that direct and chain energy losses of electric vehicles would be high; that the recharging of batteries was too slow, and that infrastructure for recharging was poorly developed. Nor did the Federal Ministry of Transport show much enthusiasm for electric vehicles. Both Ministries took the position that solutions to traffic problems should come from an efficiency revolution that would involve better use of existing roads and eliminate unnecessary traffic, such as cars looking for parking spaces in cities.

Germany's Research and Technology Ministry, however, whose responsibility it is to support the development of new industrial technologies, argued that if electric vehicles were to become a reality somewhere in the world, German firms should be there to supply the needed technology. Since

1974, the Ministry had funded new component development by the industry and had already spent 120 million DM on battery development projects. Electric vehicles as such had been built and tested for several decades in Germany, and 2,000 electric vehicles were in use in Germany in the early 1990s, most of them in fleets.

Around that time, due mainly to the California ZEV mandate, the German car industry was becoming more active in the field of electric vehicles.[33] In 1991, representatives of the Research and Technology Ministry, the Transport Ministry, industry, and several universities who visited the United States to investigate electric vehicle developments, concluded that the way to move forward with electric vehicles was to develop high-performance batteries. When it came to the criteria seen as central for users - namely range, driving performance and costs - the leading actors generally identified the battery as the key technological component that would spell success or failure of electric vehicles.

In the early 1990s, lead-acid batteries were used in most electric vehicles, but the possibilities for their further development were considered limited. Several German car manufacturers had already equipped a few cars with new high-temperature, high-energy-density sodium-sulphur and sodium-nickel-chloride batteries for demonstration purposes, and the Research and Technology Ministry itself had also spent many millions on R&D into these and other batteries. The Ministry thus decided to initiate a large-scale project to test and evaluate the latest German battery and electric drive technologies.[34]

Objectives and project organization

The Rügen electric vehicle project had several objectives.[35] The main goals were to test new battery systems and to carry out a comprehensive evaluation of both the energy and ecological aspects of electric vehicles compared to gasoline and diesel vehicles. To this end, each vehicle was equipped with a data collecting system. A subsidiary goal was to test the latest generation of electric motors. Other objectives that were officially defined as equally important but in fact played a subordinate role, included:

- A technical evaluation of fast-charging technologies;
- Evaluating the vehicles' acceptance by vehicle operators, users, and local residents;
- Proving that electric vehicles met the standards for traffic and vehicle safety;
- Setting up and operating a solar power plant to provide recharging energy for EVs; and
- Documenting the technical progress of Germany's latest electric vehicles.

Rügen Island was chosen as the location for the electric vehicle project for several reasons. Because the island is located in one of the former East German states, the project enjoyed funding from a federal programme designed to foster exchange and transfer of technology and to create employment. As Rügen is not a large island, the range of the electric vehicles would be great enough for most types of trip. In addition, as the island has many nature conservation areas and health resorts where clean, silent traffic would be advantageous, electric vehicles could create a positive public image. The 'clean' image of electric vehicles was strengthened by the fact that Rügen was already using renewable energy, mainly wind turbines, to generate electricity and, as the sunniest place in Germany, it was also suitable for the use of solar energy to recharge electric vehicles.

The project had a complex organizational structure. The R&D firm of DAUG, a joint venture between Volkswagen and Mercedes-Benz, provided day-to-day project management under the supervision of the certification and research organization TÜV Rheinland, which acted on behalf of the Research and Technology Ministry. A committee with representatives from each project partner supported DAUG. The project partners were:

- Four German car producers and one bus manufacturer;

- The manufacturers of the three different battery systems tested in the project;

- Two institutes which carried out scientific monitoring and evaluation of technical and environmental aspects;

- A joint venture of three electricity companies, two German and one Swedish, which provided the electric charging stations; and

- A subsidiary of Deutsche Aerospace that tested a new type of solar cell.

A number of local companies, institutions, administrative authorities, craft associations, and retail companies on Rügen co-operated with the project partners by testing electric vehicles in daily use. Private customers were not among the target group. The vehicle fleet grew gradually from October 1992 on but was only completed in March 1995 due to delays in supply. Participants were recruited via advertisements in the local newspaper and signed six-month vehicle leases. Use of the vehicles was free except for the electricity consumed. DAUG opened a service station for maintenance and repairs of the test vehicles, and a solar installation constructed on the roof of the station supplied some of the energy for recharging.

The Federal Research and Technology Ministry underwrote almost half of the Rügen project (26 million out of the total of 60 million DM). In addition, the State of Mecklenburg West-Pomerania added a further (300,000 DM).[36]

Rügen was the largest experimental trial of electric vehicles ever undertaken in Germany. However, the financial support the project enjoyed stands in sharp contrast to the many small dedicated EV manufacturers who received little support from the Federal government and were not invited to participate.[37] Various car manufacturers, especially Volkswagen, which converted a Golf model into the CitySTROMer, and several smaller German firms, had developed conversion electric vehicles for testing by various electric utilities and the German postal service. In addition, a number of prominent consultants, segments of the environmental movement, and some small manufacturers were producing dedicated lightweight electric vehicles with limited speed and range. However, these smaller parties, were excluded from the Rügen project and criticized the car-industry's focus on advanced batteries, longer range, and safety features. In their alternative vision, electric vehicles would fulfil specific functions, notably intra-urban travel and therefore needed to be supplemented by other transport means including public transportation and rented gasoline cars for longer distances.[38]

Overall, the number of EVs in Germany was stable at around 2,000 throughout the 1980s but included only 100 passenger cars. By mid 1994, electric vehicle stock had increased to around 4,400 (49% passenger cars, 35% vans and pick-ups, 3% buses and 13% agricultural and other vehicles). The growth was due almost exclusively to an increase in the number of passenger cars, both conversions and dedicated vehicles, stimulated by subsidies from various states (Länder) and municipalities.[39]

Project design and management

The Rügen project was managed in a professional manner by an experienced organization. As mentioned above, this was not an easy task because the project structure involved a number of companies and institutions aimed at achieving a variety of goals. Although the organizations shared most of the information gathered, they could keep certain specifics to themselves. The project progressed smoothly except for delays in the initial delivery of vehicles and limits on their availability due to product defects. The experiment was completed by summer 1996, one year behind schedule.

Also as mentioned above, actors with differing conceptions of the appropriate role of electric vehicles were excluded from the project. The conditions for participation effectively excluded small start-up companies that to date had supplied most of the electric vehicles in use in Germany. Some firms used lead-acid batteries in their vehicles and thus did not qualify for support from the Research and Technology Ministry. Other firms' fibreglass vehicles did not meet prevailing safety criteria. Such cars were considered unsafe because they did not have a crush zone to absorb some of the energy released upon impact in a collision. Some firms could not match the funding offered by the Ministry. And finally, some firms could not deliver vehicles in time because they manufactured them in small quantities.[40]

Within the project, energy consumption was bound to be high because the vehicles selected were modelled on standard gasoline-powered cars. Such heavy vehicles required excessive energy and power density from their batteries, whereas lightweight vehicles would not have required high-performance batteries to achieve acceptable performance. Although the car industry did acknowledge in principle the desirability of purpose-built, lighter-weight electric cars, for this project they wanted to concentrate on the batteries.[41]

The industry also wanted to use cars that had already been tested for safety, because safety testing new purpose-built prototypes would have added to the overall cost of the project. Critics claimed, however, that the battery systems tested on Rügen would never become available for electric vehicles because they would remain much too expensive. Among these critics was the lead-acid battery industry, which obviously had an interest in undermining support for alternatives to its products.[42]

The project managers focused considerable attention on monitoring and evaluating the experiences of users, who were surveyed at three distinct points in time. First, before they started using a vehicle, to establish their usual travel patterns for reference; second, during the test period, to register their early impressions of driving an electric vehicle; and third, after the test period, when the main focus was on whether, based on their experience, they would want to use or purchase an electric vehicle. The final report on the project was widely distributed in the form of a CD-ROM.

Government funding was essential to the Rügen project. Although the car manufacturers pointed to the money they spent on it as proof of their good intentions with respect to electric vehicles, it is unlikely that they would have initiated the experiment without the almost 50% subsidy provided by the research ministry. The industry was reluctant to support electric vehicle development and would have preferred to focus on reducing emissions from conventional cars. This pattern of behaviour shows the strong attachment of the car industry to the dominant regime and to the conventional views of its customers. In this sense, it may be more correct to speak of the car producers as 'suppliers' to the project managers rather than as true 'partners' in the project. This generalization is even truer for the battery producers.

Learning

The contribution of the Rügen project to the development of the electric vehicle niche was very limited with respect to both learning and institutional embedding.

Learning was restricted to technical issues, especially those related to the experimental battery and drive systems. The project concluded that most of the electric drives used were very safe, reliable in operation, and exhibited favourable energy-consumption patterns.[43] The technicians at one car company involved also concluded that the project had contributed to the standardization of components and to learning how to build a battery system into a car.

However, serious problems occurred with the sodium-sulphur batteries and eventually led to their withdrawal from the project in autumn 1993. On-site tests confirmed that electrotechnical flaws caused frequent breakdowns of these batteries. This fact first came to light in the early 1990s, most spectacularly so when a fire destroyed the BMW E-1 prototype, which was equipped with such a battery. Although it was a cable fault and not a defect in the battery itself that caused the fire – the battery itself went on functioning even after the fire was put out – these events undoubtedly harmed development of the sodium-sulphur battery. The companies involved in its development ceased their activities during the course of the Rügen project.

Most of the experimental results summarized the frequency and types of vehicle failures, the distances covered, and the number of days that the vehicles were actually available for users (i.e., not under repair). The charging infrastructure was used less than expected, but the project management considered the presence of recharging stations important as a kind of safety net that gave users the security of knowing they would not get stranded somewhere without power.

Little was learned about the potential market for the vehicles. The generally shared and taken for granted view was that businesses and utilities would be the early adopters of electric vehicles. The target user group in the Rügen project was representative of these users, but the findings regarding these users were very general. For instance, according to users, the vehicles' range (80-150 km without intermediate recharging) and reliability were 'sufficient'. A more surprising result was that drivers also used the electric vehicles as primary vehicles even when a conventional second car was available as a back-up.[44] However, these results were reported after drivers had been using the vehicles for only six months, so the findings may well be biased because the electric vehicles were still a novelty.

Regarding environmental aspects of electric vehicles, the Rügen project produced an unprecedented database for ecological evaluation. The evaluation showed that electric vehicles not only did not produce any noise or local emissions but also contributed less to summer smog and created less acidification on the ground and inshore waters than did conventional vehicles. On the other hand, they had a higher acidification potential and more impact on climate change because of the unfavourable mix of fuel inputs for the German electric grid. Only under the specific usage conditions of very frequent, short, stop-and-go drives did the electric vehicles score markedly higher on environmental aspects.[45]

Clearly, these results did not improve the position of electric vehicles in the environmental debate. However, the results might have been more favourable if heavy conversion-model vehicles had not been used. Nor did the project contribute to learning about new and more sustainable forms of transport and mobility. Because of the chosen target market, the test vehicles were used simply to replace conventional vehicles.

In general, learning in the project took place on a number of aspects. Technology selection was based on the assumption that the development and application of high-performance batteries was the right direction for technological innovation and was necessary if electric vehicles were to compete commercially with regular cars.

As mentioned before, the project excluded actors such as start-up manufacturers of lightweight vehicles who challenged the conventional wisdom. Nor were environmentalists who had developed visions for sustainable transport featuring lightweight electric vehicles included in the project. Users were involved in the actual testing phase but not in designing and setting up the project. Their role was passive; it was to validate or disprove the assumptions made by the original project partners. Thus, no second-order learning emerged.

Institutional embedding

Besides learning, the Rügen project also made little contribution to the institutional embedding of electric vehicles. In fact, it can be argued that beyond the immediate confines of the experiment itself, the project actually harmed the process of niche development. The manufacturers reclaimed the vehicles at the conclusion of the project and, although the project manager recommended a follow-up project, the manufacturers did not make them available for use elsewhere.

The Rügen experiment also failed to influence production plans, and it did not have a positive impact on the expectations of the partners with regard to marketability of electric vehicles. The car industry remained sceptical about the future of electric vehicles, although, in 1994, the industry, together with several electric utilities and battery producers, presented a position paper to the Federal government. The position paper estimated that a half-million electric vehicles could be in use by 2000 if the government created the right conditions, such as tax benefits, restricting access to inner cities to all but electric cars, and building an infrastructure for battery recharging and standardization.[46] The government ignored this position paper without evoking protest from the industry.

Furthermore, the project also failed to result in a stronger and more aligned network among the participating companies. The project did create additional mobility for the island's inhabitants because the electric buses were used to open a new bus line; however, that service ended with the end of the project.

Since the Rügen project, Wilfried Legat, a former leading official of the Federal Transport Ministry and later consultant to the Association of German Electric Utilities (VDEW), has made several pleas for a market-oriented follow-up project by the electric utilities and the car industry to demonstrate the everyday usefulness and reliability of current-generation electric vehicles. According to Legat, who was inspired by the La Rochelle project, the test should take place in a large city over the space of two years, and users should be 'ordinary' car drivers who would have their own recharging facilities and pay for using the cars. Several electric utilities were said to be interested in the proposal, but thus far there has been no indication that such a test project will be supported.[47,48]

A failed niche development

By 1998, the number of electric vehicles in Germany had increased to 4,500, a slight increase when compared with 1992, at the start of the Rügen experiment.[49] The car industry has abandoned battery electric vehicle development, and the Federal government does not believe in it either. Volkswagen has stopped producing the CitySTROMer. The substitution strategy has failed, although several fleet operators press on. The market niches for electric passenger cars are mainly filled with French imports and the products of small companies selling converted and lightweight vehicles. BMW and Volkswagen have shifted their attention to the introduction of hybrids. DaimlerChrysler is investing in replacing batteries with fuel cells and has repeatedly announced that they will start selling limited numbers of fuel cell vehicles in 2004.[50]

The Rügen project clearly did not contribute to further niche development and can be considered a success only if we focus on battery testing. Rügen proved that the chosen battery was not the wonder battery that many had hoped for. The learning was mainly restricted to this issue.

The strict conditions for receiving matching funds from the Research and Technology Ministry meant that only major, established actors could participate in the project. This limited participation to those whose opinions about the electric vehicle niche were already formed. The dissonant voices of small start-up firms, environmental movement adherents, and critical scientists remained outside the project. The main actors, including the Research and Technology Ministry, generally had (and still have) little faith in the vehicle quality and production capacities of the start-up firms.

The proponents of a vision of a different transportation system, one that would stand in opposition to the current dominion of the privately owned and ubiquitously used passenger car, were thus prevented from challenging the underlying assumptions of the actors who designed, organized, and carried out the Rügen project. If the Ministry of Research and Technology had included the small firms in the project, they would have gained legitimacy and might then have attracted capital and technical support from universities. In turn, these steps might have resulted in the niche development process in Germany taking a different direction. Instead, most of the small firms subsequently went bankrupt.[51]

The PIVCO experience: ecological product differentiation[52]

Introduction

Niche development in Norway was stimulated by the activities of the firm Personal Independent Vehicle Company (PIVCO), which started developing a small, purpose-built, two-seat electric vehicle in 1990. A spin-off of Bakelittfabriken, a thermoplastics manufacturer specializing in boat hulls and munitions, PIVCO was established in 1990 following the enactment of the ZEV legislation in California.

Jan-Otto Ringdal, Bakelittfabriken's managing director, initiated the development of an electric vehicle for a niche market in urban and suburban areas that would be produced in Norway. A respected and charismatic businessman, he was able to enlist the participation of some of the largest companies and research foundations in the country. These actors brought important complementary abilities to the project and gave the project status and long-term financial support. PIVCO Industries also received significant government subsidies and loans, in large part because Norway lacked a car industry. Some politicians regarded PIVCO as an opportunity to build a new industry that would compete with long-time rival Sweden. In certain respects, therefore, environmental or energy-saving reasons were secondary.

PIVCO contacted the oil company Statoil because its petrol stations could also be used as sites for recharging stations and because the company was an important actor in Norwegian industry. Other actors included the chemical and mining company Norsk Hydro, which brought its know-how in aluminium production, and electricity producer Oslo Energi, which enrolled as a possible user of electric vehicles and as an important supporter of the battery-recharging infrastructure. In addition, the national postal service sponsored PIVCO to polish its environmental image. Other companies were similarly motivated: Statoil, for instance, used the EV project to show its receptivity to new technologies and new trends in society.

Each sponsor took a minority interest in PIVCO. The largest shareholder besides Bakelittfabriken was the Statens Näringsdistrikts Kreditbank (National Fund for Regional Development and Industry), a public financial institute that supports technological development projects in Norway.[53] The Norwegian Technological Institute was responsible for testing the vehicles, partly to acquire competence in the area of electric vehicles.

PIVCO's product, developed first under the name 'City Bee' and later renamed 'TH!NK', was initially positioned as a second car for families. Conceptually, it was very similar to the Swatchmobil/Smart and was projected as a lifestyle vehicle. Based on market research data about how people use cars in urban settings, PIVCO identified the target market as follows:

Surveys show that in urban settings most cars rarely travel for more than 50 km a trip, and on average carry less than two people. A small, lightweight, electrically powered PIV (Personal Independent Vehicle) could easily fulfil this mission, and produce much less pollution in the process.[54]

This finding also corresponds with the user patterns found in an Oslo Energi survey. PIVCO thus designed and developed the City Bee with a driving range of 150 km at a constant speed (50 km/h), 110 km in city traffic, and a top speed of 90 km/h. Although the City Bee's NiCd battery was relatively sensitive to temperature, the vehicle's performance was better than that of previous generations of electric vehicles used in Norway. It was thought that this would make it more attractive for the identified user groups.

The most innovative part of City Bee was Bakelittfabriken's internationally patented thermoplastic body and aluminium space frame, which employed a process that moulded a complete vehicle body in one piece using thermoplastic material. The resulting rustproof body was coloured throughout and needed no paint, thereby obviating the need for both costly paintwork and metal tooling for body parts. An aluminium underframe achieved a structure that was satisfactorily rigid and complied with existing safety standards. The battery was placed in a separate steel case underneath the car. The choice of materials made the City Bee largely recyclable, and the investment in a manufacturing plant for this kind of vehicle was low compared to the car industry's traditional mass-production plants.[55]

This conceptual approach to vehicle production was possible because Norway did not have a tradition of automobile manufacturing. An established domestic car industry might have derailed the concept, as has happened in similar situations in other countries. Although Norsk Hydro had some experience as a manufacturer of aluminium car parts for major car manufacturers, the actors who supported PIVCO were only marginal participants in the automobile regime.

The City Bee concept was a deliberate misfit with respect to the automobile regime. Ringdal had studied other electric vehicle projects and concluded that the new company had to develop a different kind of vehicle and build it in a novel way. The City Bee, in his opinion, should be environmentally friendly not only to use but also to produce. PIVCO even referred to the City Bee as 'a vehicle for short-range transportation in cities' rather than as a 'car'. This choice of words is instructive on two counts. First, many people have negative associations with the word 'car' and, second, PIVCO did not want potential users comparing the City Bee with the traditional high-performance gasoline car.[56]

To minimize development costs and to use the best electronics and battery technologies available on the market, PIVCO decided to purchase all parts and components from independent suppliers rather than developing them itself.[57] The design concept and patented technology allowed simple, low-cost production at low volumes; the lower limit for breaking even was only 5,000 vehicles per year. The initial price estimate for the City Bee, if mass-produced, was 100,000 Norwegian crowns, although the actual price would depend to some degree on the volume produced.

The first factory was built in Norway, where production of test vehicles started in 1995.[58] PIVCO intended to produce City Bees in other countries where local joint-venture partners would, it was hoped, contribute knowledge of local markets and additional financing. Plans were made for production in California, among other places.

Because of the City Bee's technological innovations and catchy design and the strong partners in the development network, expectations for PIVCO's success were very high among both the developers and impartial observers. However, these expectations proved somewhat unrealistic after practical tests of the first prototype models revealed several shortcomings in both the design and market conceptions.

Experimental Introduction in Norway and California[59]

The first unnamed prototype was ready in 1992 and the next one, named City Bee, was on the road one year later. A series of 10 City Bees were then tested under winter conditions during the 1994 Winter Olympics in Norway. The vehicle's profile was raised further in 1995, when a City Bee won the Scandinavian Electric Car Rally, a race in stages from Gothenburg to Oslo. The next year, 100 prototypes were delivered for tests in a car-rental project in Oslo, several Norwegian fleets, and in a station car demonstration in the San Francisco, California, area.

Oslo Energi and Statoil's car rental department organized the City Bee rental project in 1996. Public interest was high, and the media followed the project closely. Six vehicles were stationed at the airport and four more at a Statoil fuel station in the centre of Oslo. The municipality and several companies joined the project and thus promoted themselves as environmentally conscious; their logos were placed on the vehicles next to Oslo Energi's, and their support was mentioned in brochures and information materials. Thirteen hotels showcased their environmental awareness by setting up charging stations, and Statoil installed charging stations at five of its fuel stations as well.

The price for renting a City Bee was set at substantially less than for a conventional car. The rental period was restricted to three days. Although many companies were interested in leasing City Bees for longer periods of as much as a year, Statoil and Oslo Energi wanted to test the EVs under the tough test conditions of an ordinary rental situation in which many different people drive the vehicles. This was seen as a way to test the market for City Bees, because customer reactions should indicate whether there was public interest in electric vehicles. Customers were asked to fill in a questionnaire about their experiences for evaluation.

From October 1995 to March 1998, PIVCO also tested the City Bee in San Francisco in a project called the Bay Area Station Car Demonstration. The purpose was to determine the viability of using electric vehicles as 'station cars' for making short everyday trips. Station cars are vehicles made available at mass-transit stations to transit riders, who can use them for any type of short trip. A household that had station cars available could eliminate the need to own, fuel, and maintain one or even more vehicles while preserving a high level of personal mobility, and, when electric vehicles are used, a form of mobility that results in minimal pollution.[60]

The station car demonstration was initiated in 1992 by the Bay Area Rapid Transit (BART) authority and Calstart. The station cars were used by commuters to reach company sites several kilometres from railway stations in Oakland. When the company that had originally committed supplying the vehicles backed out, BART invited PIVCO in as a partner at the suggestion of Calstart's CEO, who knew that the company was looking to introduce the City Bee into California.[61] PIVCO leased 40 City Bees at a cost of $20,000 for the two years of the project.[62] An added personal cost to users would have been home recharging (which would have cost less than US$1 per night), but most recharging took place at the railway station, where BART provided free electricity to station car users. However, BART found that the City Bees' small size and low speed made them too slow and unsafe, and, to reduce its liability risk during the life of the project, banned them from highway use. Drivers of the City Bee were thus restricted to destinations that could be reached via suburban and regional roads. In addition to these two projects, PIVCO also obtained important data from demonstration vehicles placed with several selected organizations in Norway, including the postal service and the electric utility companies in Stavanger and Moss.

These users monitored their experiences mainly by circulating questionnaires among the drivers and soliciting their opinions regarding the cars. PIVCO incorporated these reports, collected through regular meetings and contacts with the users, into its own internal development processes. The Technological Institute, which also took part in the evaluation process with its technological measurement equipment and laboratory tests, provided feedback to PIVCO about problems and made suggestions for improvements.

Project design and management

Although both the Oslo car rental project and the San Francisco demonstration were well designed and managed, their role in the development of the City Bee has attracted criticism. Unlike the mainstream car industry, which does not release vehicles for sale until all testing has been completed, PIVCO leased cars that were basically still prototypes to various kinds of users.

In both projects, 50 City Bees were driven only about 50,000 km. By contrast, an ordinary gasoline car model is tested more than ten to twenty times that distance and hundreds of experienced experts analyse and evaluate the test results before it is released.

PIVCO, at the time a small company with only about 50 employees, did not have access to these kinds of resources (although experienced engineers from Rolls Royce, Saab, and Lotus have since joined the company). By choosing to have real users test the prototype, the company ran the risk that any problems could damage the City Bee's reputation and undermine its future market. PIVCO admitted that its approach entailed these risks but argued that other problems could have resulted if the company had tried to market the City Bee as a finished product. The company also maintained that it was open about the fact that the City Bee was a prototype and might suffer from technical problems, and that it asked customers and users to report any problems, so they could be solved and learned from.

BART, however, found PIVCO's stance unsatisfactory and judged that the company lacked experience and expertise in car manufacturing and in understanding design, the time required for car testing, and what it really meant to develop a car. Small problems could have been eliminated if the cars had been more carefully tested in Norway before being sent to California. One representative of the Technological Institute agreed with BART. And certainly, comparing PIVCO's efforts with those made by established car manufacturers when developing a typical new car model, the conclusion is that the City Bees might not have been thoroughly tested.

According to BART and other US actors, PIVCO would have found it difficult to compete if it intended to offer electric vehicles at the same level of technology and pricing as the large automobile manufacturers, which have much greater resources and experience in serving customers. This lack of infrastructure for testing and marketing vehicles is a general barrier for outsider firms who want to enter the field of automobile production and sales.

Learning

PIVCO's practical tests were important for evaluating the City Bee's performance in everyday use, for assessing customer acceptance, and for assessing possible applications of the City Bee and EVs in general. The test also resulted in lessons regarding the station car concept and revealed various design faults. Thus, a broad learning process emerged with a high user involvement. It was, however, a learning process that clearly lacked second-order effects.

In the Oslo project, the customers were generally positive about the City Bee after having driven it. Before driving it, many customers believed that EVs were slow and heavy, and with a short driving range, but afterwards, they were usually impressed with the City Bee's fast acceleration and good driving characteristics. Most renters tended to fall into one category: well-educated men who were generally interested in new technology - for example, CEOs who rented a City Bee to drive to meetings with executives from other companies because it would show their environmental interest and awareness.

The Oslo experience changed PIVCO's market expectations. Originally the company had been targeting the second-car consumer market. But based on the Oslo test, PIVCO concluded that companies would be very important as early customers to build up this future market. Firms that wished to communicate a particular environmental image could choose the City Bee because its design was visibly different from that of ordinary cars. Several companies that have used the City Bee have pointed out the image effect and argued that the cost could be seen as part of their company's marketing budget. Other users might choose a different model; for example, the postal service would prefer a delivery van, which would fit the mail service better.

The findings on users' experiences in the Bay Area demonstration raised doubts as to whether the City Bee would be successful in the United States. Although users appreciated the City Bee's spacious interior, comfortable seats, good design, quietness, ease of parking, scratch-resistant thermoplastic body, and ease of driving, they also complained about the vehicle's short driving range, low speed, and low power capacity.

One firm that participated in the station car demonstration concluded that the City Bee might attract consumers with an interest in environmental issues, but that it would not appeal to most companies, except perhaps to city-based delivery firms. After this firm pulled out of the project due to changes in management and budget restrictions, the employees were offered the opportunity of staying involved on an individual basis, but none did. They did not want to pay for the vehicles themselves, especially because the company had started a free bus service for employees that used BART.

These experiences also indicated that the City Bee's two-seater concept was not appropriate because employees often travelled in groups of more than two when driving to meetings and lunches. To be viable in a company carpooling fleet, the City Bee would have to be a four-seater. Because the US market share for two-seaters is only 2%, promoting a car like the City Bee as a company fleet car would be difficult. Also, the short driving range was a problem for meetings held far from the worksite.

The City Bee was well suited as a station car, but its quality would need improvement and its price would have to drop if it was to be competitive after the expected release of a more price-competitive new generation of electric vehicles from the large American car manufacturers. The California users seemed to evaluate the City Bee against their expectations for a conventional car with a combustion engine and ordinary driving range and speed. However, US roads and highways are different from those in Europe and especially Norway. Americans travel long distances to their workplaces, often on highways, and shopping centres are frequently located near these highways. The City Bee's low top speed was thus a limitation for the US consumer. In Norway, on the other hand, roads are usually narrower, and there are not so many highways. The largest roads are located in and near the largest cities, and traffic density is also much lower than in California. In Norway, the City Bee was not perceived as a car, but as an electric vehicle. These factors made the City Bee much more suited to the Norwegian market than to the Californian market.

The practical tests in Oslo and California also revealed several design defects. Water leaked into the cars after rain and the fan for drying the windows did not work well, so the windows stayed blurred. Also, the brakes and suspension were weak, the windows rattled, the doors stuck, and there were problems with charging the battery and starting the motor. PIVCO, its suppliers, and the BART garage corrected these defects during the tests, and the second generation of City Bees was modified to prevent the same problems from recurring. However, the Road and Traffic Authority commented that the City Bee could not meet its normal requirements or user demands, that small electric vehicles could not protect drivers and passengers in collisions as well as ordinary cars could, and that the driving range was too short due to the batteries. Further development of batteries would be required to give the vehicles a longer driving range.

PIVCO tackled these issues in its development activities, and the latest generation of the City Bee, renamed the 'TH!NK', has been further modified according to the lessons learned in the Oslo and Bay Area tests. Among other changes, the TH!NK uses a new motor and a new battery pack, and the fully aluminium frame has been replaced with a steel lower frame and an aluminium upper frame.

The environmental effects of EVs were the subject of much discussion by the actors in these tests. One main goal of developing EV technology was to reduce pollution and solve environmental problems caused by city traffic. Environmental effects were not monitored in the Oslo rental project, but because 99% of Norway's electricity comes from hydropower, the reduction in emissions are obvious.

Although City Bee users in Norway received no additional benefits for using EVs, they did benefit from existing legislation and regulations aimed at stimulating the introduction of EVs. Norwegian owners of EVs are exempt from taxes; the Oslo municipal government waives toll fees for EVs entering the city centre; and private parking companies in the central city arranged special parking places where electric vehicles can recharge at no cost.

Whereas in Norway the City Bees were positioned as replacements for regular cars, whether household second cars or company cars, in California they were tested as a new form of transportation, the station car. BART concluded from the Bay Area demonstration project that the station car concept was functional, and, as of 1998, the demonstration project entered a new phase.

The first phase focused on testing EVs, which were the favoured type of vehicle. The next phase calls for testing a car-sharing plan, which can be seen as an elaboration of the station car plan. In the car-sharing plan, cars will be used during working hours and not just during commuting hours. For example, a first commuter can deliver his station car to the BART station, where a second commuter can use it to get to his or her worksite. Then, a third person can pick up the car at the worksite and use it as a company pool car during the day. With at least three persons using a station car - and paying for that use - in one day, the concept becomes much more financially viable. A car rental firm could take care of the rental system and the administration of the cars at BART stations.

However, BART was not planning to continue using the City Bees; rather, they were considering the Toyota Prius hybrid and the Honda Civic natural-gas car because they were cheaper, performed better, and, most importantly for a car-sharing system targeted at company pool fleets, were four-seaters.[63]

Institutional embedding

The PIVCO network in Norway was very strong and well aligned. The involvement of large companies such as Statoil, Norsk Hydro, and Oslo Energi gave the venture both status and financial stability, and attracted government funding. According to a study by Gjoen and Buland, Bakelittfabriken was initially the main actor in the development of the City Bee, with managing director Jan-Otto Ringdal serving as initiator, network builder, and promoter.[64] The other large industrial partners contributed to the development process, but were not deeply involved and did not assume any larger risks. This suggests that these actors became involved in the project more for public relations and image considerations than because of a long-term interest in EVs. The Norwegian government's role in the Oslo project was mainly that of a sponsor, but it too became involved in the development of the City Bee through statements from politicians stressing the environmental benefits of using EVs and the job-creating effects of manufacturing the City Bee in Norway.

The creation of the PIVCO network has obviously been an important impetus for the EV niche in Norway, but this has not yet been reflected in sales. As of 1999, only 300 EVs, about 70 of them City Bees, were in use throughout the country.[65] The numbers started growing in 2001 (see below).

In the United States, the Bay Area demonstration contributed to the creation of a niche for station cars and shared cars. The National Station Car Association used the results of the Bay Area demonstration, together with those of other local demonstrations elsewhere in the country, to write a proposal advocating a large national demonstration that would involve 3,000 to 5,000 cars in several cities. Also, as of 1999, BART was planning to expand the station car concept to a car-sharing system.[66]

With reference to recharging infrastructures, the Oslo tests of the City Bee included the construction of an expanded recharging network. A large number of charging facilities were, of course, already in place as ordinary electrical outlets connected to the nation's electrical grid. An infrastructure for quick recharging stations has been built up in the larger cities in Norway, where most electric vehicles are expected to be used. Maintenance networks for EVs were also established. Oslo Energi set up a maintenance service for its own EVs and those owned by the Oslo municipality, and has expressed interest in expanding to service City Bees owned by others as well, while Statoil uses the importer of Denmark's Kewet EVs for maintenance and repairs.

In the Bay Area, recharging facilities were set up at the station car parking lots near the BART stations. PG&E, the local electric company, stressed that the project would be crucial to the process of learning and developing standards for an EV recharging infrastructure in advance of the mass production of EVs in California.

The PIVCO vehicles aroused great media interest in both Norway and San Francisco, and the projects garnered many laudatory print articles and television reports. In Norway, the positive public image of the City Bee and PIVCO may have been due in large part to patriotic factors. Because there had never been a car industry in Norway before PIVCO's EV venture, there was no doubt a strong desire to see the 'adventure' succeed. Production of the City Bee/TH!NK may create positive spillover effects for Norwegian industry. The fact that Norway's king and queen were present at the opening ceremony of the Bay Area demonstration in San Francisco can only be seen as a symbol of their interest in promoting Norwegian industry abroad.

Before the Oslo and Bay Area projects, PIVCO, the companies supporting the development of the City Bee, and the Norwegian government had high expectations for both the commercial prospects and the positive environmental impact of EV technology. The two tests not only reinforced these expectations but also spread them to more actors, especially to users, despite BART's and the Road and Traffic Authority's criticisms of the City Bee's performance characteristics. The supporters of the City Bee envision that the existing technology niche will be transformed into a market niche once the City Bee is mass-produced. Nevertheless, the Norwegian actors recognize the importance of international development of the EV market for their own success. They believe that whether EV technology develops into a dominant technology will depend mainly on the development of quick-charge batteries that would extend driving range.

Ford to the rescue

The tests of the City Bee resulted in PIVCO deciding to start series production of the TH!NK in Norway, but in 1998 the company ran into financial problems that forced it to look for external financing. Before then, PIVCO's owners, the Norwegian government, and a stock offering had financed the development work.[67] In October 1998, the company opened a new production plant and also unveiled the TH!NK vehicle at the 15th Electric Vehicle Symposium in Brussels. The vehicle received widespread acclaim from the automotive and environmental press, and PIVCO used the exhibition to seek new investors or partners but was forced to declare bankruptcy a short time later. Nevertheless, in January 1999 Ford Motor Co., in a surprise move, acquired a majority interest in the company.

For Ford, acquiring the interest in TH!NK Nordic A/S (the company's new name) offered several benefits. It suddenly had a vehicle that satisfied the ZEV mandate and could be sold in California.[68] Also, the TH!NK would give Ford international entrée to the market niche for ultra small cars. Moreover, the TH!NK would give Ford experience in small-scale, flexible production. As Ford president and CEO Jac Nasser commented:

This car not only will give us immediate access to a whole new market niche, it will provide a wealth of new ideas for us to develop. We are particularly interested in new concepts in the use of plastic body components, as well as low-volume and flexible manufacturing.[69]

When Ford stepped in, PIVCO had reportedly received as many as 500 orders for the TH!NK from banks, energy companies, and the Norwegian postal service. In July 1999, the Norwegian telecommunications firm Telenor reportedly placed an order for 600 to 700 vehicles.[70] By the end of 1999, TH!NK Nordic had begun production of the new model, which was to be launched first in Norway, where it would be delivered through Hertz, Ford's car rental subsidiary. Lessees would receive discounts on other rental vehicles when they needed a larger car for family vacations and other activities, and Hertz would provide a replacement vehicle when the TH!NK required servicing. Furthermore, Hertz was to include the EV in its rental car line-up. Marketing in the rest of Europe and in the US started in 2001, in time to meet the deadlines set in the ZEV mandate. The initial sales target was set at 5,000 vehicles a year.[71] If the company meets this target, the TH!NK will become the best-selling battery-EV worldwide.

La Rochelle: market differentiation[72]

Introduction

In France, electric vehicles have been the subject of attention several times in recent decades. In the 1970s, the state-owned electric utility EDF attempted to commercialize EVs after fuel cell research became sidetracked. Fuel cells, which had first been used in space flight, were the subject of several national research programmes aimed at developing them for electricity generation, but EDF favoured the nuclear-power option. The focus then shifted to other possible applications, the most promising of which appeared to be electric traction. In the end, after the failure to develop fuel-cell-powered electric vehicles, EDF decided to promote battery electric vehicles.

Because EDF saw this technology as a good opportunity to sell more electricity and perhaps indirectly legitimize nuclear power, it led the research. At the time, however, industry, and France's automobile producers in particular, were unwilling to commit themselves to the technology. Both Renault and Peugeot-Citroën S.A. (PSA) restricted themselves to developing and testing several prototypes that were electrified versions of existing models of small passenger and company cars.[73] By the end of the 1980s, about 500 electric vehicles were in use in France. Fifty of these were experimental vehicles built by Renault and PSA, while the others came from several smaller firms.[74] These firms can be divided into three groups: dedicated EV manufacturers that produced mainly vans; producers of voiturettes (described earlier in this Chapter); and producers of specialized utility vehicles. From a strategic perspective, Renault did not choose to push electric vehicles, while PSA opted for a first mover approach.

The first sign of real commitment to EVs on the part of PSA came in the 1980s when the company created a special research team dedicated to EVs, separate from its existing research department. This separation came about for two reasons: first, because of the specific competencies required by EVs, and, second, because of senior management's concerns 'not to trouble' the rest of the company with EVs. The team worked directly under top management, to whom it regularly presented prototypes, thereby validating the quality of its work. The relationship of trust that developed between the researchers and management seemed to hold out the real possibility that PSA's EV programme could succeed.[75]

In the early 1980s, the team began developing plans for a small fleet of electric versions of the 205 model. The researchers considered outside funding necessary if PSA's management was to be convinced of the viability of their undertaking, so they began seeking partners for the project. They found them in EDF and the municipality of La Rochelle, which thus became involved in testing the vehicles. EDF was an obvious choice as the utility already had extensive experience with EV technology, and La Rochelle was already the site of EV testing by EDF and by the municipality's fleet operator. In addition, the municipality had been pressing PSA for a project that would generate employment in the town to compensate for the shutdown of PSA's Talbot factory several years earlier. Together, the three partners successfully applied for EC funding, which marked the start of a long-term co-operative effort that has continued until the present day.

The 1980s saw tests of the electric 205 model and other prototypes and the first attempts to commercialize electric vans. Several hundred were built, and EDF bought a large number for use in its fleet. EDF played an important role because it was not just a customer but also a partner that could give valuable feedback to PSA.

By the end of the decade, PSA had learned much about EV technology from its experience with the vans and was convinced that the time had come to implement its concept for a small electric passenger car. PSA believed that it had achieved the best possible integration of the electric drive system and was using the best available battery for the intended market (nickel-cadmium). A strategy for the 1990s was defined that was to start with commercialization of electric vans for use in urban fleets followed by the mass production and marketing of conversion-design passenger cars for use in fleets and by private individuals. Subsequent stages would introduce a purpose-built electric car – several concept cars and prototypes have been developed to this effect – and hybrid and fuel cell cars.

The decision to commercialize the conversion-design EVs would depend on the success of the La Rochelle experiment, which was designed to answer one remaining question: would private users accept this electric vehicle?

Objectives and project organization

The La Rochelle experiment, which involved practical testing of 50 EVs by real users under daily operating conditions over a period of two years, is in a way the conclusion of many years of increasingly close co-operation between PSA, EDF, and the Municipality of La Rochelle. The experiment started in December 1993 with 25 Peugeot 106s and 25 Citroën AXs,[76] which were leased at 1,000 and 900 French francs a month respectively to private users and companies.

The objectives were to demonstrate the EVs' usefulness and performance as well as to study users' reactions, the technical viability of EVs for daily use in urban areas, and the recharging modes. PSA trained its dealers in the area to maintain the vehicles. EDF installed the necessary infrastructure for kerbside recharging, at private parking lots and at service stations. The municipality provided incentives for EV use by reserving special parking places for them.

The users, all volunteers, were recruited through advertisements in the local media and direct mailings from PSA dealers to their customers.[77] PSA had defined the potential target customers as users of second cars who lived in households with more than one vehicle, drove less than 10,000 km a year, mainly in cities and seldom or never on highways, did not use their car for holidays or weekend trips, and made very few trips of more than 100 km. This potential group of users was estimated at 1-2% of the total market. In the actual experiment, however, the user group was biased toward professional men, employees, and senior executives, while retired people were under-represented. In addition, the cars that were replaced by the EVs were not always second cars or cars that belonged to the low-range segment.[78]

To gain feedback about users' experience, the participants were monitored extensively during the initial eighteen-month period. The three partners focused their monitoring and analysis on learning about EV driving, use patterns and recharging behaviours, the development of a perceived relationship between users and the vehicle, the evolution of the EV's status and image, and the integration of EVs into the management of travelling needs, such as trip planning. After the initial eighteen months, PSA decided to extend the experiment a further six months and to replace the prototypes with series-produced vehicles that had been slightly modified to take user feedback into account.

Evaluation of project design and management

As the result of two years of preparation, the La Rochelle experiment was very well organized. A steering committee was established with different groups in charge of different tasks such as public relations, contact with users, feedback on vehicles, maintenance, and charging stations. This organization benefited from the strong commitment of the three partners, who shared a firm belief in the prospects of EVs. Their commitment was reinforced by the fact that the partners had already been co-operating for a long time. Also important was the fact that the partners saw the experiment as part of a step-by-step process for introducing technological innovation. Over the course of the experiment, new technological elements were incorporated one by one, the idea being not to come up with a futuristic car whose development would have been uncertain because too many technical problems had to be solved at the same time. This also explains why PSA chose to convert existing cars into EVs rather than building purpose-designed ones.

The location, size, and duration of the project also deserve mention. The choice of location was the object of intense discussion within PSA, which had to choose between Versailles, which was near the research team's worksite, and La Rochelle, where earlier experiments had already taken place. Versailles had the benefit of proximity, which had proved crucial during the 205 project when PSA engineers had to travel to La Rochelle each time there was a technical problem, but Versailles lacked experience with EVs. La Rochelle was 400 km from the EV team's worksite, but the municipality had the major advantage of long-term experience with EVs and a record of co-operation with the EV team. In the end, the arguments in favour of La Rochelle were decisive.

Another point worth considering is the choice of a medium-sized town such as La Rochelle versus a larger city such as Paris or Lyon, where it is likely that electric vehicles would find a larger market. One might assume that driver behaviour would not differ radically but, clearly, battery-recharging behaviour would be different due to the greater distances in the larger cities. Also, unlike medium-sized towns such as La Rochelle, large cities usually have fewer houses with private parking spaces where users could recharge their EV from sockets at home. This would mean that larger cities would have more need of kerbside recharging facilities and in public parking places, which in turn would require a larger financial investment.

With 50 vehicles and a somewhat larger number of drivers, the user sample was large enough for the reactions of a given target group – owners of small second cars – to be assessed. Had the aim been to evaluate overall potential demand, a larger-scale experiment would certainly have been necessary. Nevertheless, the combination of low rental prices and the cost of maintaining the 50 vehicles was already quite an investment in terms of money and human capital.

Finally, the La Rochelle experiment lasted two years including the final six months during which series-produced vehicles were tested. As mentioned above, one aim of the first phase was to improve production EVs based on user feedback and to analyse users' perceptions of EVs. After changes were made to the electric vehicles in the wake of these evaluations, the resulting series-produced vehicles were tested in the second phase to see whether the changes affected users' perceptions. Because the experiment grew out of a long period of research and development during which most technical problems had already been solved, the duration of the experiment appears to have been adequate.

Learning[79]

PSA set up the La Rochelle experiment to learn about user needs for a particular market niche. It was less interested in gathering information about design and other issues, although it did implement design changes as a result of the experiment. The main issue was whether enough people were willing to buy the small electric passenger cars that PSA had been developing since the early 1980s. PSA considered user involvement crucial to the development of its EVs, so the process was targeted to produce first-order learning, that is, testing user acceptance of a specific vehicle. What did PSA find?

The general level of user satisfaction turned out to be high, and perceptions of the EV were positive, with limited driving range being the only drawback identified by the participants. As an unintended consequence, however, the experiment also resulted in second-order learning effects, for a new product identity took shape in the course of the experiment.

Users developed a new relationship to the EV in three phases. First, they discovered the electric vehicle and found that it was a 'real' car that was pleasant to drive because of its silence, comfort, and cleanliness. Second came the stage of 'maturity', during which both the advantages and restrictions of the EV became clearer. This led users to modify travel planning by giving up long trips, avoiding random travel, and using the EV specifically as an urban vehicle. In the third phase, users came to define their vehicles as a different kind of car. For example, they began using the EVs mainly for short trips and their regular car for long trips. Also, household members tended to share use of the EVs more easily than was typically the case for traditional cars. By the end of the experiment, most users routinely recharged their vehicles at home. At first, users would precisely control the context in which their EV was used; many treated their EV too cautiously, recharging every day even when it was not necessary. Later, as they began to trust the EV more, they recharged less often.

The partners were satisfied with the technical choices that had been made during the development process, and the EVs exhibited few technical defects.

However, users pointed out that some of the EV's specific functions were underdeveloped. For example, a number of elements, such as the reverse button and the charging cable, came in for particular criticism and were modified before series production began. Half of the users who took part in the experiment decided to buy the EV they had tested, a clear indication of their satisfaction with and enthusiasm for EVs.

The experiment also yielded feedback regarding the battery chargers installed in and around La Rochelle. EDF proposed that ten changes be made to the normal chargers and 14 to the fast chargers. The project findings showed that private users who owned a private parking place preferred to recharge at home, and that locating chargers near office buildings, public places, and restaurants increased their use. EDF also observed that users would need public charging places for psychological reasons - even drivers who recharged at home would find it reassuring to have other stations available.

Institutional embedding

The three partners' efforts in La Rochelle since the early 1980s created a niche for the development of EVs. The 50-car experiment contributed to the development of this niche by confirming initial expectations about the vehicle's technology, yielding lessons about its conditions of use, and strengthening the alignment of the network. The network developed was broad, including a variety of parties, but it lacked outsider involvement. PSA dominated the network, so that the interests of the car regime were protected. With a strong environmental movement lacking in France, no actor in the experiment had the objective of energy-efficiency high on the agenda, as EDF produces 'CO$_2$-free' electricity in abundance from its nuclear plants.

The technical performance and reliability of the electric Peugeot 106 and Citroën AX led PSA to invest in small-scale production of these vehicles. Production began in November 1995, and three years later, 30 electric vehicles were being produced each day in the factory of Heuliez/France Design, a specialized small-scale car manufacturer.

Production numbers were expected to increase when sales grew, but sales remained far below projected results. The initial hope was to sell 2,000 cars per year, rising to 10,000 annually by the year 2000. However, actual sales were 1,300 in 1996 and just 800 in 1997, despite significant subsidies to EV buyers from both the national government and EDF. A PSA executive hypothesized that the 1996 production satisfied pent-up demand for the commercially produced EVs. In 1998 and 1999, sales increased again and by late 1999 about 6,000 battery electric vehicles, 80% of them PSA products, were on the road in France.[80]

These figures conceal a hidden shift. By the end of the period, almost the entire demand was from fleet customers. PSA sold scarcely any of its EVs to the individual customers that had been targeted in the La Rochelle experiment. This is apparently partly the result of the implementation of a 1997 air-pollution law that empowered towns to restrict traffic to clean cars permanently if needed, and required government agencies, public bodies, and some large companies to increase the proportion of their fleet vehicles running on electric power, LPG or natural gas to 20%. This law supported the market niche for electric delivery cars, so in 1998 PSA acted on this demand by starting production of the Peugeot Partner and Citroën Berlingo delivery cars, based on the 106 and AX used in Rochelle.

Although PSA's sales have not met projections, Joseph Beretta, head of the EV group, commented:

> Our EVs may not make a profit – but they don't make a loss either . . . We have to earn money from our involvement with electric vehicles – so far we have invested around one billion FFr in electric- and hybrid-vehicle development. We currently have around 80% of the electric vehicle business in France, and we intend to remain the dominant company in this sector.[81]

In addition to deciding to start series production and the interest among test users in buying the EV that they had used, a third indication of institutional embedding resulting from the La Rochelle project was the huge interest it attracted among the public. Although the partners organized an active and intensive public relations campaign around the experiment, the extent of public interest in the project still surprised them. Many people visited La Rochelle and the project received wide media coverage, which increased the use of and interest in EVs in France. In addition, the EV users in La Rochelle have also strongly supported EVs locally and have formed a user club that has played an important role in promoting the technology. Similar clubs were also created in other cities.

Since the end of the La Rochelle experiment, PSA has become involved in three major programmes or projects that can be considered as branches of the EV niche developed in La Rochelle. The first is Vedelic, a research programme in the Poitou-Charentes region, where La Rochelle is located, aimed at developing a version of the Peugeot 106 model with lithium-carbon batteries. This project was developed to promote economic development in the region, to develop high-level industrial and university research, and to protect the environment. The region is seen as a strong contender for developing an innovative EV industry because many relevant research institutes and industries, including car and electronics companies, are already located there. Several government bodies are providing most of the project funding.

A second project involved testing a plan for integrating rental EVs and scooters into the multimodal public transport system that was developed in La Rochelle since starting in the early 1980s. This project aimed to test the feasibility of such a system and to identify types of use and potential customers, such as tourists, citizens interested in buying an electric vehicle, or households that occasionally need a third car. The rental system scheme was extended with another experiment developed by the Liselec group, a partnership between PSA, a transport operator that was in charge of the overall design of the system, and a third company that is providing computer management and telematics. PSA is the largest contributor to the project, which started in March 1997 with 10 vehicles being tested on PSA sites. A follow-up test commenced in 1998.

The Liselec experiment aimed to provide self-service electric vehicles that could be accessed with smartcards. Five stations equipped with the necessary recharging infrastructure were developed with five electric vehicles each (Peugeot 106 or Citroën Saxo, the successor of the AX model). The Liselec experiment is a step toward the development of the more innovative TULIP concept, which is comparable to the Praxitèle concept discussed in Chapter 5. The TULIP concept is a service that gives subscribers access to small two-seater EVs located in stations throughout the city. A central control station manages the system and handles booking, maintenance, and payments of fares. The stations are set up with a computerized parking space management application and an automatic recharging system.

Third, Peugeot in England was one of the main partners in the 1996-1997 Coventry Electric Vehicle Project, in which 14 Peugeot 106 Electric cars and mini-vans replaced existing gasoline and diesel vehicles used by five organizations in the English Midlands. The project assessed the extent to which EVs could reduce urban air pollution, overall energy consumption, and emissions of greenhouse gases from the transport sector, as well evaluating the on-road costs of the 106E under real operating conditions.[82]

Peugeot had several reasons for participating, including market-testing of EVs in the UK, increasing its 'green credentials', and forging links with government bodies and fleet operators. According to one PSA executive, the main aim was to sensitize fleet operators to the potential benefits of electric vehicles.

Frustrated electric urges

Although the La Rochelle experiment was a success, it failed to produce the hoped for result – fast-growing sales of the 106 and AX to private consumers as a second car for city use. Several explanations for this failure have been suggested, including PSA dealers' lack of commitment to selling EVs, a lack of marketing effort,[83] and the possibility that PSA simply overestimated the market on the basis of too small an experiment and underestimated the positive bias that coloured the views of those involved in the actual experiment.

All these factors are certainly valid, but we would like to propose another explanation. The users' positive assessments only emerged after living with the EVs for a relatively long time. This suggests that a second-order learning process occurred in which users gradually rethought how they used the car. Over time, users learned to recognize and appreciate some of the car's specific and interesting features, and changed their mobility patterns accordingly. However, this process was not captured by PSA, which did not make use of the 'relationship' that developed among the users and was nurtured in user clubs. Instead, the company relied on its own assumptions about why users liked the cars. Evidence that contradicted the company's own beliefs about what the EV drivers liked and needed – for example, in the case of the charging stations – was not properly interpreted.

In short, the EV turned out to be more radical for users than PSA wanted it to be. The company had designed a vehicle that it thought would fit easily into the existing mobility pattern of a specific group of users. Because that fit turned out not to be present, selling EVs to private customers requires more effort than selling regular cars. One conclusion that we have drawn from the La Rochelle experiment is that, because EV users needed to become convinced of the vehicles' advantages, using the vehicle for a longer period is a prerequisite.

The La Rochelle experiment led to a process of niche branching by contributing to the development of a niche for delivery cars. The initial project also was important in supporting decisions to develop more advanced individualized public transport systems. Praxitèle, one such plan that incorporated EVs, will be discussed in the next Chapter, which focuses on reconfiguring mobility. In certain respects, the PSA EV product strategy anticipated the introduction of hybrid cars (and fuel cell cars). PSA's experimental hybrid cars use some of the technology, such as the electric motor and the electronic controls of the 106, developed for battery-powered vehicles. In 2000, PSA launched the Berlingo delivery car as a hybrid using a small auxiliary LPG engine to extend its range.[84] Also, Citroën introduced a passenger hybrid car in 2001.

Mendrisio: new mobility option[85]

Introduction

The Mendrisio experiment, which focused on the contact and competency network for so-called lightweight electric vehicles (LEVs) in Switzerland, was discussed earlier as an example of an outsider bottom-up approach.

This network emerged in the second half of the 1980s as an effort to develop 'solar cars'. Before long, a technological trajectory of LEVs developed, and some noteworthy technological breakthroughs in lightweight vehicle construction were achieved. In the early 1990s, the Swiss Federal Office of Energy set up a promotion programme for LEVs aimed at consolidating the network and moving the most outstanding and promising prototypes into the manufacturing stage. Concomitantly with the technological network, a protomarket developed in which pioneering consumers tested and supported the development of LEVs.

The Mendrisio experiment aimed not only at expanding this market niche but also focused on further exploring a new mobility system in which the use of LEVs would lead to new mobility patterns that would integrate various transport means for various purposes.

Surveys of LEV users showed that the majority was men aged 35 to 54.[86] The LEVs' high purchase price meant that their owners needed an above-average income; the early users were also more highly educated. The LEV users' mobility patterns differed greatly from what would have been predicted on the basis of their demographic profile. The LEVs were driven an average of 18 km/day or 3,000 km/year, a distance well below the maximum range for one battery charge. Many users needed an individual vehicle for commuting or business-related trips, and they often lived in small rural communities where public transport was infrequent. Most of them first learned about LEVs by visiting automobile expositions or reading about them in the media. A third of the users had participated actively or passively in the Tour de Sol, but few had actually seen LEVs on the street before buying one.

In most of the user households, the gasoline or diesel car remained the principal car, but the LEV was used for some of the travel that would otherwise have been made with the regular car or by public transport. Twenty per cent of the LEV users had the LEV as their only vehicle, and some of these replaced their gasoline car with it entirely. Overall, LEVs did not appear to increase the users' individual mobility but rather led to a decrease of between a quarter and a third in total kilometres travelled.

Both the surveys and interviews with LEV users showed that they had adapted their mobility patterns to the new technology. The LEV's limited driving range meant that users chose them for short individual trips and used their gasoline car or public transport for long trips, family trips, and for transporting heavy goods. The range restrictions meant that drivers started planning trips more carefully by looking for the shortest routes and avoiding energy-intensive trips. LEV users also applied their heightened consciousness of energy usage to domains other than transport and reported instituting energy conservation measures in their households. Furthermore, LEV users learned to drive more cautiously; their relative rarity, and their silent operation, meant that bicyclists and pedestrians could fail to notice them. Also, drivers of regular cars often underestimated the LEVs' performance and would engage in dangerous overtaking manoeuvres.

The LEV users criticized the vehicles' high purchase price, limited range, battery maintenance problems, unsuitability in winter conditions, inadequate safety features, and relative discomfort. On the other hand, they expressed satisfaction with LEVs' speed, acceleration, reliability, and ease and cost of maintenance. Although half of the users preferred an EV that resembled a conventional car, the other half expressed a preference for a car that was distinctively identifiable as an LEV. Also, most users said they experienced an increase in driving quality and appreciated the lower speeds because they contributed to a more relaxed lifestyle. Summarizing these findings, Harms and Truffer concluded that:

> In total, LEVs can be considered as a means of leading to a 'disarmament in traffic' and more conscious mobility patterns.[87]

Users reported that the most important factor in their buying decision was the LEV's environmental friendliness, followed by the vehicle's suitability to urban traffic conditions. Quite unimportant were the status conveyed by driving an LEV or the idea that buying an LEV would support their further development. Although all users considered the environmental aspects a key issue in their purchase, the surveys revealed different emphases on other issues. Harms and Truffer[88] distinguished four types of LEV users:

1   'Eco-promoters' whose overriding concerns are ecological and political issues and who buy LEVs as an ecological alternative to the gasoline car.

2   'Techno-promoters' who are more interested than other users in technological issues and view LEVs as a technological product that deserves support.

3   'Individual urbanites' who are primarily interested in alternative mobility and find LEVs an interesting new form of urban mobility.

4   'Affluent inquisitives' who are motivated first by curiosity, like the idea of owning an exclusive vehicle that is different from normal cars, and can afford to buy and maintain one.

Objectives and project organization

Given the inadequate capacity for manufacturing prototypes and the slowdown in the early market for LEVs, in summer 1995 the Swiss Federal Office of Energy (Bundesamt für Energie – BFE) decided to carry out a large-scale experiment with LEVs. The goals of the project were multi-levelled. First, it was conceived as a market acceptance project that might put pressure

no

on vehicle developers to improve their product and considerably cut production costs. A second focus was to evaluate policies for the wider diffusion of LEVs all over Switzerland. And finally, the project hoped to focus on how LEVs could become part of an innovative overall strategy for the development of new intermodal mobility patterns.

In a qualitative sense, the experiment's main goal was to engender the diffusion of LEVs in a particular community in order to create a laboratory situation for testing new transport systems. The quantitative target was to motivate a community of 10,000 inhabitants to replace about 8%, or 350, of their cars with LEVs within six years. At that point, the community would have become the sought-after laboratory or microcosm in which the role of LEVs in new mobility forms could be analysed and simulated.

The idea of such a large-scale test first emerged from a workshop organized by the BFE in 1991. The workshop aimed to bring all the main actors in the LEV milieu together to decide on a governmental promotion programme. During the workshop, the idea was not given high priority because it would have consumed considerable resources that, at the time, it was felt could be better invested in improving components and prototypes. In 1993, however, after the significant market slowdown, one workshop participant re-introduced the idea at the annual LEV conference and proposed a detailed project plan. BFE became interested in the presentation and provided financial resources to carry out a feasibility study in 1993.[89]

The diffusion of LEVs was to be supported by a series of promotional measures, and the main goal of the study was to judge the effectiveness of these measures, which would indicate how a politically supported process for diffusing LEVs throughout the country could be achieved. The goal set was to replace 8%, or 200,000, of the standard vehicles in Switzerland with LEVs by the year 2010. Among the wide variety of measures tested in the project, the most notable was a large subsidy on the purchase price of an LEV depending on its energy efficiency. LEVs and electric bicycles received up to a 50% subsidy, and other EVs received a 30% subsidy.

In addition to this main goal, the experiment aimed at analysing the environmental impact of LEVs and their influence on users' mobility patterns. The central question was whether LEVs could form an essential element of new, more sustainable transportation patterns. Other goals included:

1   Demonstrating that the advantages of EVs outweighed their disadvantages. Both policy-makers and the public were to be convinced of the viability and everyday usefulness of LEVs. Also, a strong infrastructure for maintenance and battery recharging was to be created.

2   Evaluating marketing and market introduction strategies to discover factors other than price that might determine buying decisions.

3   Studying the effects of EV use on traffic patterns, including whether users' driving patterns changed after they started using EVs, and whether they drove more or less.

4   Testing a concept of mobility that distinguished between short-distance transport means such as LEVs, public transport, and bicycles and long-distance means including public transport and rental cars. The project was to offer LEV users additional mobility services including car sharing, car rental, taxi use, and public transport.

5   Studying the environmental effects and energy consumption of LEVs under everyday conditions.

One of the most remarkable aspects of the Swiss experiment was the process by which the test community was chosen. The BFE issued a public call for a community willing to host the field test. Thirty-four communities – many more than expected – from all across the country applied, and five were invited to submit a more detailed proposal (see Figure 4.2). These detailed proposals were all of high quality, which made the choice difficult, but in December 1994, the community of Mendrisio, near the Italian border in the canton of Ticino, was chosen.

One major factor that influenced the decision was the region's broad support for the experiment, because the canton, the community, local garages, and other major actors actively supported the project, and over time the

Figure 4.2 The communities that applied in the call for tender

project management proved to work professionally. A weak point was the low environmental consciousness among Mendrisio's population compared to other regions. Before the experiment, only a few LEVs had been sold in Ticino. This made Mendrisio a risky choice because the researchers hoped to generalize their results to other parts of the country. Also, Mendrisio's location near the Italian border suggested that there might be problems in extrapolating the lessons learned there to the rest of Switzerland. The cultural specifics of the region meant that the motivations of early LEV-users in the German-speaking parts of the country were probably not comparable to those in Mendrisio. Nevertheless, the commitment evidenced in the community's proposal outweighed the seeming idiosyncrasies of the region, and Mendrisio was selected.

Project design and management

The testing project started in Mendrisio in June 1995. The community's limited size, about 6,500 inhabitants in the town and 20,000 in the neighbouring communities, allowed the inclusion of a number of partner communities. The latter were important if the experimental results were to be transferred to all Switzerland's different language regions because measures that were not suitable for Mendrisio could be tested among the partner communities. As Urs Muntwyler, head of the LEV promotional programme, put it:

> One test community is not enough to make relevant statements about the users' behaviour. The attitude of people in the German-, French- or Italian-speaking parts of Switzerland is too different. Therefore the test pattern has been extended to 'partner-communities' in all parts of Switzerland. This makes the 'Grossversuch mit Leicht-Elektromobilen' (the large-scale test with lightweight electric vehicles) an ideal field for market research for all of Europe.[90]

The partner communities carried out their own pilot and demonstration projects, which were directly connected to the Mendrisio experiment. They also received administrative, financial, and technical support from the BFE. The Federal government also subsidized the purchase of LEVs under the same conditions as in Mendrisio (27% of the purchase price), but due to budget limitations no more than 10 vehicles could be subsidized per community per year. The types of LEVs sold in the partner communities depended on the activities of local car dealers. Usually the same vehicles were sold as in Mendrisio, but sometimes only a few types of LEVs were offered. All the partner communities also introduced additional support measures for LEVs.

From the very beginning, special attention was given to the selection of the kinds of vehicles to be offered in the large-scale experiment. All types of EV were admitted as test vehicles: electrically assisted bicycles, motorcycles, three-wheelers, private cars, pickups, buses, and lorries. However, the level of the subsidy offered depended on the energy efficiency of the vehicle involved. In addition, a number of 'comparison vehicles', including efficient gasoline-driven or hybrid vehicles as well as EV prototypes not yet available on the market, were included in the experiment.

One important aspect of the experiment was the financial contribution of both consumers and producers, as reflected in the division of costs among the parties. The experiment cost about 33 million SFr, with 65% (21.5 million SFr) spent on buying vehicles and 9% (2 million SFr) on building the recharging infrastructure. The remaining money was reserved for project management and associated research. As the organizer of the experiment, the Federal government was legally restricted to paying no more than one-third of the total costs. It was users who bore the biggest share (43%) of the costs by actually purchasing the vehicles. Firms that sold vehicles had committed themselves to cutting their prices by about 10%. Furthermore, private sponsors, the canton, and the community guaranteed the remaining 17% of the budget.[91] Also, the experiment was accompanied by a careful monitoring system and about 8% of the total costs were earmarked for research. As Schwegler stated:

> The large scale test [. . .] is not only a pilot and demonstration project but also a research project. However, it is no conventional, theoretical research plan, but an uncommon, large-scale practical test on a scale of 1:1.[92]

The project plan required that the large-scale test be evaluated every two years and that a decision then be made as to whether and how the project should be continued. The evaluation after the first period was largely positive. The experiment had received broad support from individuals, firms, and organizations at the local, regional, and national levels, indicating a strong basis for continuing the project. By the end of the first phase, 82 vehicles had been sold in Mendrisio and more than 200 in the total project, including the partner communities. The initial bestsellers were the converted voiturettes and City Els because their Swiss importers were able to start offering them in Mendrisio at short notice. More established car manufacturers, however, considered Italian-speaking Ticino a relatively difficult market because it would entail extra costs such as translation costs and were therefore more reluctant to offer their EVs through their importers in Mendrisio. Nevertheless, as time went on, several manufacturers became interested in testing their vehicles in the project. And after several large companies' electric four-seaters, including the Peugeot 106, the Citroën AX, and the Volkswagen Golf began being offered, demand shifted to these models.

The project management was happy to be able to include these vehicles because the voiturettes and three-wheelers were considered to be of poor quality; also, the established manufacturers were able to provide better after-sales service. On the other hand, this made the project management reluctant to involve pioneer firms in Mendrisio. The management felt that the experiment had to remain a highly competitive platform in which the presence of some twenty different types of vehicles would allow serious comparisons of the quality of each model:

> We conduct technical tests. We measure the energy consumption. And then we see: [. . . This lightweight vehicle] isn't as good as they always say. It's miserable.[93]

The participation of smaller firms was also blocked after the first year when the project management decided that firms that could not give the 10% price reduction would be excluded from receiving subsidies.[94] This was a major problem for developers and small firms that were already operating under tight financial limits. Some of these firms thus became more interested in offering their LEVs to the partner communities, where the 10% discount was not a requirement but the subsidies were still in effect. Besides, people in the partner communities were more familiar with LEVs than was the population of Mendrisio, and this greater environmental awareness made possible buyers more interested in alternative vehicles.

Several modifications were made to the project in the second phase, which ran from July 1997 to June 1999. These included expanding the recharging infrastructure, improving the co-operation between Mendrisio and the partner communities, improving the supply of information to users, interested citizens and garage owners, setting up LEV user clubs, exempting EVs from taxes, developing new mobility concepts by combining LEV ownership with membership in car-sharing projects and annual passes on public transport, extending LEV and battery guarantees to three years, and starting to plan for the period after the test ended.[95] The project management also tightened the energy-consumption limits for receiving purchase subsidies.[96]

The enactment of several of these measures led to a boost in sales of electric vehicles, as shown in Figure 4.3.

The best-selling vehicles were two-wheelers, and the Peugeot 106s made up half of the four-wheel vehicles. The main reasons for the boom in vehicle sales in Mendrisio are the following.

- The Peugeot garages developed a new leasing scheme for batteries, bringing the purchase price below a 'psychological' limit which makes the LEVs seem inexpensive. Furthermore, PSA extended the warranty of the batteries to 4 years.

- The number of communities entitled to funding was increased in

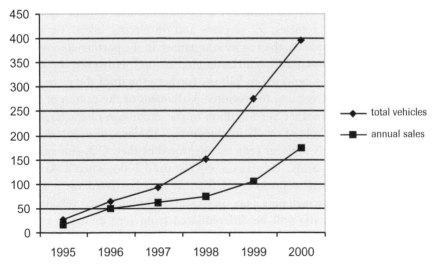

Figure 4.3 Electric vehicles in circulation in Mendrisio (Sources: www.infovel.ch and Blum and Wegmann [see note 97])

accordance with the market areas of the participating garages. These had criticized the unequal treatment of their customers until then.

• The subsidies were increased from 50% to 60% of the vehicle price. It is not clear how important this was because the general population knew little about the exact subsidies, but it may have persuaded the interested but not yet decided customers.

• Apparently there is a lead-time for decision-making which had been underestimated at the beginning. Many interested customers decided only after considerable time and inquiries to buy an LEV.

• Experiences of customers with LEVs were largely positive. The project manager was able to handle problems without creating much negative public attention.

• In general, the quality and reliability of the vehicles increased considerably in the last few years.

• Among the two-wheelers scooters had the most important share. Their sales grew in proportion to the sales of gasoline scooters in the general market. Apparently the difference between electric and gasoline scooters is perceived as acceptable. Also, scooter salesmen were much more aggressive than passenger car dealers were.

The boost in sales continued in 2000, and the project management even had to freeze the subsidies because the year's budget was already exhausted by May. By then, the target of 350 electric vehicles had been reached. In January

2001 the project had helped bring 396 EVs onto the roads, made up of 174 'cars', 29 light duty vehicles, 97 scooters and 96 electric bikes. Two-thirds were owned by individuals, the rest by companies. In the partner communities there were an additional 250 vehicles in circulation.[97] However, the lack of additional subsidies is expected to halt the further growth of the niche.

In its Spring 2001 session, the regional parliament of the canton of Ticino decided to allocate another SFr 6 million to the expansion of the large-scale experiment to the whole of the Ticino territory.[98] The new project 'vel 2' will pay subsidies to highly efficient vehicles (measured by their CO_2-emissions per kilometre) such as hybrids. This set-up will last until 2004, when it should be complemented by a new project 'vel 3'; this will be based on a general bonus/malus-system which puts a tax on CO_2 and reimburses the vehicles with higher efficiency. Tariffs will be differentiated from zero emission up to a maximum of 120 g CO_2/km, which corresponds to approximately 6 litres gasoline/100 km. The Federal government's response to these plans is generally quite good. The subsidies had been heavily debated both at the canton level and the federal level, but ultimately, when continuation was accepted by the parliament of the canton of Ticino, the Federal Office of Energy and the Federal Office for the Environment also granted funding. Additional financial support is expected from large private companies such as Electricité de France.

Learning

The learning process in Mendrisio was broad and aimed at second-order learning. Although the latter did not occur, a broad range of individual results with regard to specific technological characteristics of LEVs did emerge. One crucial problem was the erratic performance characteristics of the batteries, especially in extreme situations. Second, factory defects and concerns with the overall quality of the LEVs were identified. Third, energy use, especially by the imported French voiturettes, turned out to be much higher than expected in everyday use.[99] These technological setbacks did not lead to a sustained effort to upgrade the vehicles or to the hoped-for development of industrial LEV production. Instead, they resulted in the removal of LEVs and a change to using converted vehicles from established car manufacturers. Thus, opportunities for mutual interaction among technological choices, demand, and possible incentives were neglected. The Mendrisio experiment changed from being a test designed to explore a new mobility regime to a field test for the market introduction of various EVs. This led to a more conservative, car-oriented route of technological development, rather than the more expansive outcome that had originally been envisioned.

Major lessons were learned regarding the effectiveness of the promotion measures.[100] A survey in Switzerland's three major language regions revealed that the promotion measures were widely accepted. About 80% of respondents in the German- and French-speaking regions were in favour of promoting LEVs, as were fully 93% in the Italian-speaking regions. With regard to individual promotion measures, the subsidies were considered by far the most important. Respondents' awareness of the promotion measures varied greatly. Infrastructure measures such as guaranteed accident recovery services and reserved parking lots for LEVs were cited by 70–90% of LEV users in Mendrisio. Only about half of the users were aware of additional measures such as lower fees for renting cars or free driving courses for LEVs. A survey to gauge awareness of the various promotion measures at the end of 1995 revealed relatively little knowledge among the population of Mendrisio; at that time, LEV users judged the vehicle purchase subsidy as most important, while indirect measures to foster the emergence of new forms of mobility were judged less important.[101] A repeat survey six months later made it obvious that InfoVEL, the project's administrative office, had done its work well during the interim: knowledge had increased significantly within the population.

Of special importance were the learning processes at the user level. The reductions in mobility and the emergence of conscious mobility planning, which had occurred in the early LEV market niche, did not occur. In 1997, a detailed survey of mobility behaviour found that 64% of LEV users (including cars, motorcycles, and bicycles taken together) replaced another vehicle with their LEV, while 36% bought it as an additional vehicle.[102] With regard to mobility behaviour, 62% said that they had not experienced any change in their mobility needs since buying the LEV, while 15% reported an increase in mobility and another 15% a decline. They reported little use of promotion measures, such as free memberships in car-sharing associations, car-rental discounts, or public transport memberships, aimed at supporting shifts in mobility behaviour. However, public transport options were much less developed in the Mendrisio region than in other parts of the country.

What accounts for these results? As stated earlier, users in Mendrisio differed from early LEV users in that their environmental consciousness was much lower. For them, the car was an important lifestyle component and status symbol. Mendrisio has about 650 cars per thousand inhabitants compared to 450 cars per thousand inhabitants in the German-speaking regions where LEVs were first used. The average distance driven per month by LEVs in Mendrisio was much greater than in the rest of the Swiss LEV niche.[103] This can be attributed to the greater mobility of the population in Mendrisio. However, it might also be seen partly as a by-product of the experiment, because more converted cars ended up being driven, and these performed much better than the lightweight vehicles offered at the outset of the LEV niche.

The most important factor, however, may be that the experiment's promoters, in their public strategies, actively referred to 'normal' cars in the hope that this would enhance LEV sales, thereby inflating expected mobility. Also, local garages were involved in LEV sales activities.[104] Thus, both the marketing messages and the distribution channels locked consumers into the gasoline performance ideal, which in turn created expectations that did not fit the LEVs. Consequently, most users were interested in buying converted vehicles, which offered the same comfort as normal cars, rather than small, lightweight vehicles.

The first private LEV users in Mendrisio were often motivated to purchase their vehicle because of the local authorities' enthusiasm about the large-scale test. Because these buyers realized that the success of the experiment would be decisive for the future of LEVs in Switzerland, they wanted to contribute actively to its success. Other buyers were curious about the new vehicles and wanted to buy them to test out the new technology. These early users were satisfied with their vehicle,[105] but no separate data on their mobility behaviour are available.[106]

In the partner communities, large numbers of users purchased vehicles that differed from conventional cars, not only visually but also in performance. These users may have had different motivations from the Mendrisio users, but the better sales of small LEVs might also have been due to the differential activities of local car dealers.

The experience gained in the early LEV market niche confirmed the electric utilities' belief that an extensive battery-recharging infrastructure was not needed because most users charged batteries at home or at their workplace. However, the utilities did provide charging facilities at places where people could charge while shopping because this availability was seen as important for psychological reasons. This was also held out as the reason for developing a charging infrastructure for the whole country that would be subsidized through the government's LEV promotion programme.

The Association of Swiss electricity companies (VSE) opted for the simple Park+Charge system, which consisted of parking spaces next to lockers that contain electric sockets, a special label on the vehicle, and an energy card. Before charging, users validate the card by writing the date on the card, which must be displayed on the windscreen during charging so the police can see it. Users pay for the label, for renting the key to the locker, and for the energy card.

Institutional embedding

By the end of 1999, the large-scale test had helped to introduce 275 vehicles in Mendrisio. In 1999 sales increased dramatically due to revised incentive schemes (increase in subsidies from 50% to 60%) and new products (bikes and scooters) and services (battery leasing) in the market.

At the start, the experiment was supported by a dense – i.e., broad and highly aligned – network of people, firms, and institutions at the national as well as regional level. This network changed over the course of the experiment. The large-scale experiment was built on the experiences and networks that had emerged in the pioneer days of the Swiss LEV initiatives, but only a few of the retail and service firms from that original network actively participated in the experiment. As was mentioned above, developers and small firms found it difficult to qualify for subsidies and were effectively excluded from the experiment after the first year.[107] Quality problems were also stated as a reason for not including the pioneering firms more actively.

Before the experiment began, several established automobile firms were surveyed about their attitude towards such an undertaking. Among European firms, PSA, Renault, and Volkswagen showed interest in the project but were also sceptical about whether Mendrisio was the best site. As Claude Cygan, manager for EVs at Citroën Suisse, said:

> A number of our dealers are interested in getting to know LEV technology. They [. . .] want to introduce LEVs in their region. They are aware that they first have to invest much time and money without immediate returns. To date, the market volume was too small for broad marketing [or] advertisement campaigns.[108]

Thus, although car manufacturers were involved, they were less than strongly committed to the experiment's success. By the end of 1999, about 100 of the EVs in use were converted gasoline cars. Half were Peugeot 106s, but Honda, VW, Citroën, Renault, Fiat, and Daihatsu had also sold vehicles, and Toyota started selling the hybrid Prius during 2000. Daimler Benz planned to introduce its electric A class to Mendrisio as well, but the project was cancelled, in the wake of the merger with Chrysler.

In the end, the Mendrisio experiment suffered from a sub-optimal network because it could neither draw on the indigenous competence available in the national LEV milieu nor obtain serious commitments from the established automobile industry. Expectations were high at the start of the experiment. However, when sales failed to grow as quickly as expected, and EV prices in general did not fall as predicted, expectations of the future became more uncertain. Material embedding occurred because of the building of maintenance facilities and the charging infrastructure in Mendrisio, but these developments are subject to reversal.

Unexpected niche branching

The most promising developments related to Mendrisio occurred outside the experiment. From 1996 on, the LEV sales rate began increasing again after several years of stagnation, and annual sales nationwide increased to nearly 400 vehicles. Most of these consisted of new classes of vehicles such as electric bikes, electric scooters, and small three-wheeled pedal-electric vehicles. The increase in sales was partly due to the support measures in place in Mendrisio and the partner communities, but most of the new sales took place in other areas. One explanation is that the LEV retailers were more committed to selling LEVs than regular car dealers were, and that they offered not only a product and a service but also a 'sense of family'; this is discussed in more detail in the Twike example below.

Moreover, bikes and smaller vehicles are cheaper and perfectly usable in cities, where ease of parking and manoeuvrability on crowded streets are more important than range and speed.[109] Most power-assisted bikes can achieve speeds of about 20 km/h using just the electric motor, and some are capable of considerably higher speeds. The first power-assisted bikes entered the Swiss market in 1994, but sales took off when several lower-priced models (as low as 2,000 SFr) were introduced in 1996 and 1997. The most promising was the Swiss-designed BK-Tech Flyer, developed with funding from the LEV promotion programme; 230 were sold in the first year. By 1997, as many as 10 manufacturers were offering power-assisted bikes. Manufacturers hoped that these would become a new trend that would revive the shrinking bicycle market.[110]

Figure 4.4 Photograph of a Twike

In addition, the Swiss-manufactured ultralight (250 kg) Twike had, after years of development, reached a level of technical performance that made it an interesting product even without support measures. It is a three-wheeled two-seater that has an electric drive combined with bicycle pedals so that the electric motor can be supplemented with muscle power. More than 600 have been sold since the foundation of the Twike AG in 1992.[111] And because the Twike was admitted to the Mendrisio test only in autumn 1998, purchase subsidies did not play a role in many of the sales. The Twike bears little resemblance to a normal car (see Figure 4.4). In addition to its alternative drive system, it is steered by cable, has pedal brakes, and lacks any interior decoration. It is very efficient; its energy consumption was just 7.4 kWh/100 km as measured in the large-scale test, which corresponds to less than one litre of gasoline per 100 km. However, it has a top speed of 85 km/h and a driving range of 40 to 80 km at constant 50 km/h, depending on the size of the NiCd battery. The price is about 24,000 SFr, and an all-electric version is also available.

Because of its good driving characteristics, innovative technology, low energy consumption, outstanding design, but inadequate comfort, *MobilE* described the Twike as 'an ideal vehicle for individualists and pioneers'.[112] Both the company and its clients can be seen as the heirs to the pioneering LEV network. They are actively advocating a new philosophy of passenger transport in which the Twike would be the privately owned 'first' vehicle and its owners would share cars and public transport for other trips. The nation-wide car-sharing organization Mobility offers memberships to Twike owners for only 60 SFr a year.[113]

One example of this pioneering attitude was the 1998 'Twike Challenge', a journey of a half-dozen Twikes from Switzerland to the North Cape and back in daily stages that averaged 200 km. According to *LEV-News:*

They also succeeded in conveying the enthusiasm of the new driving philosophy to the public. First, as a new way of travelling that [recalls] the coach age: rolling along in a leisurely one-hour rhythm with charging stops, at an average of 40 km/h, with much time to watch the changing landscape and weather. Second, as series of encounters, [for] driving a Twike immensely fosters social contacts. Everyone from farmers' wives to mayors (and unfortunately, a few thieves as well) wanted to find out about the vehicle. Third, and of course not least, as a message from Switzerland that non-polluting means of transport exist . . . [that] they perform well, and are useful for many applications. They are even safe, as proven by two Twike accidents [in which . . .] both the pilots and co-pilots got off lightly . . . Now we hope that the Twike can motivate many people to take on the personal 'challenges' of using this vehicle instead of a gasoline vehicle every day.[114]

The 'Twike network' has also been a force behind the further development of the recharging infrastructure in Switzerland, Austria, and Germany. The Twike and similar vehicles can use ordinary electrical sockets. Home charging is adequate for most trips, but trips over longer distances will require recharging on the way. When the Twike company asked its customers whether they would be willing to make their home or garage available to other Twike users, over half agreed. No fees were involved because it was a service 'among friends'. The company then compiled a directory of the volunteers and sent it out to customers. To make the recharging breaks enjoyable as well as useful, the directory included information about nearby restaurants and places of interest. This simple idea of providing a recharging infrastructure, dubbed the 'Charge and Drink' concept, was then extended to all LEV users through the formation of the Working Group Recharging Stations LEMnet, made up of the Twike company, the Elektromobile Club Schweiz, the Park+Charge organization and *MobilE* magazine. In co-operation with several other organizations in the countries mentioned, and with financial support from the Swiss Energy Office, the group has compiled a guide to more than 500 sites in Switzerland and neighbouring countries where LEVs can be recharged.[115]

Both the Flyer electric bike and the Twike were indirect outcomes of the original large-scale, LEV promotion programme, which can thus be said to have been successful with respect to market development. Further joint ventures in the LEV industry may strengthen the supply side of this market.

Lessons from the experiments

The four experiments compared: some basic figures

The PIVCO, Rügen, and La Rochelle experiments were all of about the same scale, i.e., covering between 50 and 100 vehicles. This number is adequate if only a few vehicle types are being tested and if a restricted number of questions (experiences with the station car model, tests of components, user acceptance) are at issue in the experiment. At first glance, Mendrisio differs from this pattern because it was conceived as an impartial field test for major producers of EVs worldwide. Also, the goal was to stimulate new mobility alternatives and to create a realistic, if small-scale, diffusion path in the chosen community. Nevertheless, if we take the actual number of car-like vehicles - 80 four-wheelers – sold in Mendrisio during the experiment, Mendrisio is comparable to the other experiments.

With regard to funding, the experiments show variability as well. Rügen was the most expensive especially on a per vehicle basis (see Table 4.3); almost half a million Euro were spent per vehicle there. In La Rochelle, costs reached 150,000 Euro per vehicle, while Mendrisio cost about 100,000 Euro (calculating according to the official target of 350 vehicles set in circulation), and the station car project cost only 30,000 Euro per vehicle in external funding. The same order holds if the duration of the experiment, measured in vehicle-years, is taken into account as well, but it is noteworthy that Rügen and La Rochelle are more similar than the other two projects.

When considering government contributions to the project costs on a per-vehicles basis, it appears that Rügen had both the highest proportion and the largest absolute contribution. Mendrisio received 33% of its financing from

Table 4.2 Comparison of project parameters of the four EV experiments

Project	Technology	Period, costs, follow-up	Objectives	Partners
1. EV project on Rügen Island, Germany	60 EVs (passenger cars, minibuses, mini-vans, medium-sized buses), three different advanced battery systems	*Oct. 1992–Oct. 1996 *Costs: 30 million Euro (industry: 16 million, ministry: 13 million, Land Mecklenburg: 0.2 million) *No follow-up despite proposals by DAUG and electric utilities	Testing of battery and drive systems; ecological and energy evaluation; evaluation of fast-charge technology; evaluation of acceptance; demonstrating safety of electric cars; operation of solar plant	5 car companies; 4 battery manufacturers; 2 scientific institutes (monitoring); DAUG (project manager); TÜV Rheinland (project supervisor); Federal Ministry of Science and Technology
2. PIVCO EV development, including car rental project Oslo and station cars project in California	Purpose-built two-seat EVs on aluminium space frame with thermoplastic body (100 vehicles)	*PIVCO founded in 1991; Oslo car-rental project, May–Nov 1996; US station cars Oct. 1995–March 1998 *Costs: total development costs, 30 million Euro; station car project, 0.7 million from state govt, 0.4 million US govt, 0.1 million others *Follow-up: unclear; Ford did announce start of production	Building a new kind of transportation vehicle for short trips in cities according to a new concept with recyclable plastic and aluminium; helping to reduce environmental problems	PIVCO owned by Bakelittfabriken (plastics technology), Oslo Energy (utility), Norwegian Post, oil company Statoil, and a state-owned bank; other partners, Hydro Aluminium (supplier) and R&D institutions; Ford bought PIVCO in 1999
3. EV experiment in La Rochelle, France	50 EVs (passenger cars and minivans), recharging stations (normal and fast-charge)	*Dec. 1993–Dec. 1995 *Costs: PSA spent 9.1 million Euro *Follow-up: new experiments (Vedelic, Autoplus, Liselec, Coventry EV Project) and commercialization of electric vehicles by PSA	Analysis of technical behaviour of vehicles in real use by individual and professional customers; use of recharging outlets; behaviour and satisfaction of drivers; demonstration of existence of a market	Car producer PSA (vehicles); state electric utility EDF (recharging facilities); La Rochelle municipality (host, incentives)
4. Large-scale trial of lightweight EVs in Mendrisio, Switzerland	EVs (passenger cars; 350 planned), recharging stations (normal and fast-charge)	*Start June 1995, five-year duration *Costs: 23.5 million Euro (33% from federal government, 43% from user purchases of EVs, 10% from car importers, plus funding from sponsors, cantons and municipality) *No follow-up planned	Replacing 8% of vehicles in Mendrisio with EVs through more than 50 promotional measures; monitoring energy consumption; evaluating marketing strategies; assessing role of EVs in new mobility system; ultimate goal: 8% EVs throughout Switzerland.	Federal Energy Agency (funding); municipality (host); research institutes (evaluation);car importers; electric utility (recharging) Canton; partner communities; sponsors; project manager, InfoVEL centre in Mendrisio

the Federal government, while the station car experiment received about 90% of its funding from the regional government and the US Ministry of Defense. The La Rochelle experiment was funded entirely by the market partners.

These simple comparisons should not be taken as direct measures of the projects' relative efficiency. Project costs depend on both the goals and the set-up of a project, and, as a rule, technology development projects are more expensive than mere market-acceptance experiments. Furthermore, the vehicle types and thus costs are quite different. The lower cost ratios in Switzerland, for instance, are partly due to a larger proportion of less expensive vehicles. Furthermore, to calculate total project costs on a per vehicle basis is not entirely correct because every project includes numerous activities that are independent of the number of vehicles tested. Experiments that aim at simply gathering data will, for instance, cost less per vehicle than more ambitious projects. However, the table indicates the extent of financial resources associated with specific experimental designs. A final cost-benefit analysis can be carried out only if all the benefits of these projects, as well as the costs, have been identified.

Table 4.3 Cost per vehicle ratios for the four experiments

	Rügen	Pivco (in California)	La Rochelle	Mendrisio
Total project costs (in kEuro)	30,000	1,200	9,100	24,000
Total project costs per vehicle (in kEuro)	500	30	180	70
Total project costs per vehicle-year (in kEuro)	125	13	90	15
Government share of costs	45%	90%	0	33%
Government contribution per vehicle (in kEuro)	225	27	0	40

Contributions to niche development

Through the Rügen experiment, German car manufacturers learned that the sodium-sulphur battery was not the wonder battery they had been hoping for. A practical test was necessary to find this out, since all laboratory work needed had been completed. The experiment taught the partners much about electrical components, knowledge that has turned out to be useful for the hybrid and fuel cell vehicles now in development. From this perspective, Rügen can be counted as a success because important lessons were learned.

At the same time, it might be said that the project was an expensive way to learn these lessons. Opportunities were missed, as a result of the ambivalent attitude toward electric drive on the part of the major manufacturers. Many participants in the development network were involved only to hedge their bets in case electric power actually succeeded in the market (i.e., as a result of committed steps like the ZEV mandate and the sustained efforts of competitors like PSA in France). In choosing to locate the test in a protected space on an island, they also chose a prison, because it would be difficult to take the next step - for example, toward city use. The project did not support a vision of itself as an explicit experiment, nor were users adequately involved. Because institutional embedding was not a focus, niche development did not occur. On the contrary, one could conclude that the experiment had a negative effect in that it confirmed the preconceptions of the car industry and national government agencies regarding the inability of EVs to replace standard cars. The results were that funds for further EV development in Germany were blocked, networks failed to develop, and small firms had difficulty surviving. Consequently, EV developments connected to new mobility forms received no support.

From a Strategic Niche Management (SNM) perspective, Rügen was clearly a case of misplaced investment in a 'wonder battery' that neglected promising learning processes and factors that could have supported institutional embedding.

PIVCO obtained important knowledge about its EV design through its experiments in Norway and California. This was a case of advanced marketing that used prototypes to gain access to user demands. It led to a new car, the TH!NK, and a set of competencies that became important for Ford. The company concluded from the experiments that end-consumer use was not the best target market, and changed its focus to company fleets.

The experimental projects contributed to institutional embedding through the beginning of production, the creation of a generalized awareness of EVs for future markets, especially in Oslo and Norway, and the building of a strong network. Although this network turned out to lack the financial resources to move into production, and Ford had to come to the rescue, this does not diminish the importance of what was accomplished. PIVCO was an interesting acquisition precisely because of the experience it had gained and partially embodied in a new design.

From an SNM perspective, PIVCO was a case of the successful development of a new kind of product. It shows the importance and advantages of user involvement in marketing such radical new products. It also shows, however, that second-order learning and the development of new mobility patterns do not emerge automatically. PIVCO leapfrogged in design but was conservative in changing user preferences. Effort is also called for to develop learning processes on the demand side. The network involved in PIVCO was an outsider network on the supply side; it missed outsiders on the demand side, which, coupled with a radical new design, might also have pushed for experimenting with new kinds of mobility patterns.

The La Rochelle experiment was a small miracle in that selected users loved their converted electric vehicles. The project was a success in many ways. The project partners learned much about user acceptance and the conditions needed to support it. A number of design flaws also came to the fore. The public relations effort surrounding the project created tremendous awareness that led to similar experiments in other French cities, and a strong, highly aligned network emerged. Finally, a set of national level policies developed that provided EVs with special benefits in urban areas and forced fleet owners to integrate alternative vehicles, including EVs, into their fleets.

The miracle turned out to be elusive, however. Far fewer consumers than expected were willing to buy the new car outside the experimental setting. PSA was therefore forced to focus on the fleet market, where it was able to sell enough vans and delivery cars to sustain its efforts. We attribute this lack of success to PSA's failure to understand the reasons why participants actually liked the EV – that is, they liked the EVs precisely because they had features that differentiated the EV from a standard gasoline car. Users crafted their own meaning out of having to plan their mobility differently, but this message was not taken up in the sales and marketing efforts that followed the original experiment. To do so would have required PSA to work around existing vehicle distribution channels where comparisons between EVs and gasoline cars were inevitable. To succeed, such an effort would have necessitated the creation of new opportunities for consumers to experience the novel features of electric driving, for example through leasing or renting schemes. PSA might have considered retaining the experimental setting to give users the feeling that they were part of an avant-garde.

From a SNM perspective, PSA thus failed to open up to second-order learning and to act upon lessons learned by involving the users more in the learning process. PSA could have worked with emerging EV user clubs, but the company did not take advantage of these opportunities, perhaps because of PSA's entrenched position in the transport regime and its initial focus, which was to build new markets for second cars that could replace existing cars. In other words, PSA's strategy was to generate the need for an additional household car.

At first sight, Mendrisio looks like an ideal SNM case in that the design involved partner communities. The call for volunteer communities created considerable public attention and high expectations in Switzerland, and thus proved a good starting point for communicating activities regarding the experiment. The experiment aimed at a far-reaching learning process that included learning how to change mobility patterns and how to mass produce the radical new designs embodied in the LEVs. The project also intended to contribute to institutional embedding and encourage diffusion. Other aims included evaluating a wide array of incentives and supporting measures, which would provide a basis for further policies.

Something went wrong, however. The experiment turned into a market test for conservative designs, i.e, converted vehicles. Production of LEVs failed to occur, networks in Switzerland became more fragmented because the pre-existing LEV network was excluded from the experiment, and users in Mendrisio did not experience any changes in their mobility patterns. However, it would be unfair and premature to conclude that the experiment was a failure. The project is still continuing, and unexpected developments, encouraged partly by developments outside the experiment in the LEV market niche, may yet occur. The Twike fad is a case in point, because this new type of LEV has, in fact, resulted in new mobility patterns. Mendrisio also resulted in many important lessons, for example, about supporting measures and user acceptance of a variety of vehicles.

From an SNM perspective, Mendrisio revealed a dilemma for the actors involved in that emphasizing early success in terms of sales and diffusion led in this case to pressures that forced the experiment down an easier, more conservative path. However, it is clear that to attract sponsors, concrete sales targets are needed.

Lessons for Strategic Niche Management

These four cases clearly show the importance of experiments as the only way to test radical designs in practice. Real-world experiences always generate results different from studies in controlled laboratory settings. However, the projects also show that experiments are not enough to promote wider diffusion of radical new designs. A minimum set of incentives, for example, subsidies, benefits, and mandates for fleets, was needed to encourage users to experiment.

The cases also show the importance of carrying out a variety of experiments. SNM policies should not aim at one big experiment only but rather should encourage a number of experiments to flourish. This diversity will allow the emergence of surprises and unexpected lessons. Examples include the Twike niche-branching in Switzerland and the rapid growth of the EV fleet market in France.

Processes of niche development are nonlinear and hard to predict. New expectations will arise, experiences gained will influence choices, and wider developments may exert an influence as well. In the case of battery-powered electric vehicles, for example, the ultimate relaxation of California's ZEV mandate was important because it dimmed market prospects for EVs. In general, it is difficult to explore a proposed new regime through experiments because there is often pressure to follow a conservative path. Pressures may come from a need to register success in the short term, as well as from the involvement of actors who are a part of the existing regime. Outsiders, on the other hand, may find themselves excluded from the network. Thus, SNM policies that aim at regime exploration must manage the network's membership, which will influence the range of space available for exploration of new user preferences. This, however, brings with it a second dilemma in that the network involved in the experiment needs resources such as production competencies and financing that often come from traditional actors. Bringing together actors and expectations is a major element in SNM.

Second-order learning, though important for initiating co-evolutionary dynamics, can be difficult to achieve. It is important not to focus too much upon shortcomings such as limited range, high price, or poor detailing. Instead, emphasizing positive new features can be used to create a new product identity. For this, new distribution channels and active lead users are essential. In this sense, setting up user clubs and other platforms for exchange between users would seem to be a central element for SNM.

Electrifying mobility revisited

The supportive political environment for battery-powered vehicles has faded in the last few years. The progress made in battery development has not been as spectacular as hoped for, and the attractive model EVs presented by the automobile industry have in most cases remained only prototypes. Furthermore, political attention has shifted away from environmental concerns and energy conservation to problems of economic growth and stability. Finally, and most importantly, new technologies have emerged to supersede those that even recently seemed to hold great promise.

One indication of these changed conditions was the 1996 amendments to the California ZEV mandate. As mentioned above, that year CARB eliminated ZEV requirements prior to the 2003 model year because battery technology was not considered adequate to warrant widespread introduction of EVs. CARB started discussions about including other clean vehicle technologies such as hybrids in the ZEV regulation, which could give manufacturers partial credits. This would acknowledge that some cars can have lower emissions than battery-powered vehicles when the pollution caused by the generation of electricity is taken into account. Two years later, CARB decided to allow auto manufacturers to meet a portion of their ZEV requirements with gasoline-powered vehicles or hybrid vehicles that produced emissions in the so-called 'super-low-emission-vehicle' (SULEV) category. Several manufacturers have already demonstrated such cars.[116] Hybrid-electric vehicles also fall into this category.

The largest car companies are also developing fuel cell vehicles, which, as pure electric vehicles without the energy density restrictions of battery-powered vehicles, also qualify as Zero Emission Vehicles. Fuel cells are currently seen by many as the ultimate clean and efficient energy source for automotive and other applications. Starting in the early 1990s, several car manufacturers supported intensive fuel cell R&D activities, in particular the so-called Proton Exchange Membrane fuel cell.[117] The leading developer of fuel cell vehicles is an alliance between the Canadian firm Ballard Power Systems, DaimlerChrysler, and Ford Motor Company. DaimlerChrysler has demonstrated several fuel cell vehicles under the name NECAR ('New Electric Car'), and believes that it will have a fuel cell passenger car ready for commercialization by the year 2004 and a bus by 2003. While various car manufacturers seem to be willing to settle for buying fuel cells from the Ballard alliance when they eventually become available, others are developing their own fuel cell vehicles in-house or in partnership with other firms.[118]

The prevalence of SULEVs, hybrids and fuel cell vehicles is understandable from our discussion of the characteristics of the private transport regime in Chapter 3. These vehicles fit into this regime more easily than BEVs: although they may have revolutionary technology under the hood, the way their users need to handle them and the performance they get is not very different from regular gasoline cars.

Battery-powered vehicles are currently mainly considered for use in business fleets and car-sharing systems, in combination with information and telecommunication technologies, and in business fleets. This pattern is also emerging in other countries. However, it would be wrong to rule out battery-powered EVs for other applications, because future battery-powered vehicles may be developed and used in unexpected ways. As was argued in the previous Chapter, battery-powered vehicles, more than hybrids and fuel cell vehicles, allow for the development and exploration of a new mobility regime. This seems to have already occurred in Switzerland, where the Twike quickly gained popularity as an adventurous and sporty vehicle. Technology and user requirements co-evolve, and policy must take this into account by keeping future options open.

Notes

1   This chapter makes extensive use of Hoogma, R. (2000) *Exploiting Technological Niches. Strategies for Experimental Introduction of Electric Vehicles*, Ph.D. Thesis. Enschede: Twente University Press.
2   This section is based on several sources, mainly Mom, G. and Van der Vinne, V. (1995) *De elektro-auto: een paard van Troje?* Deventer: Kluwer Voertuigtechniek; Kirsch, D.A. (1996)

The Electric Car and the Burden of History: Studies in Automotive Systems Rivalry in America, 1890–1996. Unpublished PhD Thesis, Department of History, Stanford University, San Francisco; Mom, G. (1997) Geschiedenis van de Auto van Morgen. Cultuur en techniek van de elektrische auto. PhD Thesis, Technical University of Eindhoven, Eindhoven; Fogelberg, H. (1997) The Electric Car Controversy: a social-constructivist interpretation of the California zero-emission vehicle mandate. Chalmers University of Technology, Göteborg; and Velde, R. Te (1997) Van Technologische Niche tot Dominant Regime. De verspreiding van de benzineauto (1859–1906). Discussion paper, University of Twente, Enschede. We also profited from discussions with Gijs Mom, David Kirsch and Frank Geels.

3   See Thompson, F.M.L. (1976) Nineteenth-century horse sense. *Economic History Review*, **29**(1), pp. 60–81; and Tarr, J.A. (1996) *The Search for the Ultimate Sink. Urban Pollution in Historical Perspective*. Akron, Ohio: The University of Akron Press, chapter 12.

4   Quotation taken from Mom and Van der Vinne, *op. cit.*, p. 127. See also Nye, D. (1990) *Electrifying America*. Cambridge, MA: MIT Press, for this enthusiasm about electricity and its many applications.

5   Fogelberg, *op. cit.*

6   For that reason they were banned in many American cities. In England, they became subject to very strict regulation (one of which was the requirement that a man with a red flag walk in front of the vehicle – the 1865 Red Flag Act) which made diffusion virtually impossible. See Flink, J. (1993, 1988) *The Automobile Age*. Cambridge, MA: MIT Press, p. 2.

7   It was at one of those high-speed races that a French electric rocket-shaped vehicle by the name of '*la jamais contente*', achieved the speed record of more than 100 km/hour. This event is often referred to as proof of the good performance of electric vehicles in those days, but such extreme performance would wear out the electric race vehicles very quickly (Mom, *op. cit.*, p. 114).

8   Hård, M. and Knie, A. (1993) The ruler of the game: the defining power of the standard automobile, in *The Car and Its Environments. The Past, Present and Future of the Motorcar in Europe*. Proceedings from the COST A4 Workshop in Trondheim, Norway, have called this combination of attributes the paradigm of the 'race and travel limousine', which became the exemplar for the technological trajectory of individual mobility for the rest of the twentieth century.

9   Abt, D. (1998) *Die Erklärung der Technikgeschichte des Elektromobils*. Frankfurt/Main: Peter Lang, p. 332.

10   In this context Mom and Van der Vinne, op. cit., speak of the electric vehicle as a 'Trojan horse'. See also Kirsch, *op. cit.*, pp. 70–71.

11   Flink, *op. cit.*

12   Mom, *op cit.*, pp. 139–165 and 296–342.

13   Fogelberg, *op. cit.*, p. 40.

14   See Mom and Van der Vinne, *op. cit.* These European EVs comprised 1,700 in Germany (including 863 passenger cars, 554 company cars and 275 three-wheeled vehicles), 318 in France, followed by Switzerland with 200 and the Netherlands with 115 (70 passenger cars, 38 company cars and 7 three-wheelers.) Based on statistics from an issue of the *Allgemeine Automobil Zeitung*, no. 39, 1915.

15   Mom, *op cit.*, pp. 554 and 565.

16   Wakefield, E.H. (1994) *History of the Electric Automobile: battery-only powered cars*. Warrendale: Society of Automotive Engineers, p. 259.

17   In fact, CARB forced the introduction of cars that fell in four progressively more stringent categories of vehicles: transitional low-emission vehicles (TLEV), low-emission vehicles (LEV), ultra low-emission vehicles (ULEV), and zero-emission vehicles (ZEV). Manufacturers were allowed to produce any combination of the first three groups as long as they met increasingly stringent fleet average emission requirements each year. The best source of information on this regulation is the website http://www.arb.ca.gov/msprog/zevprog.

18   For example, positive crankcase ventilation to reduce hydrocarbon discharge and exhaust control devices to reduce emissions of carbon monoxide, nitrogen oxides and lead compounds became mandatory in the state in the 1960s, two decades earlier than Europe.

19   Schot, Hoogma and Elzen, *op cit.*

20   In the ZEUS project, which was completed in June 2000, 1200 zero and low-emission vehicles were introduced in the eight cities. See www.zeus-europe.org. For Citelec, see www.citelec.org.

21   CARB aimed at having 400,000 ZEVs on the roads in 2003 and expected that by the year 2010, 70% of all vehicles in southern California would be electric. This was only considered possible if the car industry were to change to producing electric cars. (Interviews with CARB officials for the University of Twente study *'Towards Cleaner Cars and Transport. Country Study USA'.* Report for the Dutch Ministry of Transport and Public Works, September 1995.)

22   An inside account of the history of the Impact/EV-1 is given by Schnayerson, Michael (1996) *The Car That Could. The inside story of GM's revolutionary electric vehicle.* New York: Random House. The controversy about the California zero-emission vehicle mandate is analysed by Fogelberg, *op. cit.*

23   Hoogma, R. (1999) Report of mission to Japan for the INTEPOL project, 24 November – 13 December 1998. University of Twente, Enschede.

24   Schwegler, U. (1999) Über 200 E-Fahrzeuge für Postzustellungen in Frankreich. *MobilE*, No. 2, pp. 14–15.

25   A German firm took over the production when the Danish producer went bankrupt in 1995 (*Stromthemen* 10/96). Another Danish company has produced over 800 electric two-seaters since 1991.

26   See Truffer, B. and Dürrenberger, G. (1997) Outsider initiatives in the reconstruction of the car: the case of lightweight vehicle milieus in Switzerland. *Science, Technology and Human Values*, **22**(2), pp. 207–234.

27   Truffer and Dürrenberger, *op. cit.*

28   VSAI (1997) *Neuzulassungen von Personenwagen 1980–1997.* Bern: Vereinigung Schweizerischer Autoimporteure. (http://www.vsai.ch)

29   For a discussion and comparison of both types of design see AVERE (1992) *A Systems Analysis Study on Advanced Electric Drive Systems for Buses, Vans and Passenger Cars to Reduce Pollution.* Brussels: AVERE.

30   This equals, for instance, the amount of energy per person-kilometre needed for public transport in Switzerland. Synergo (1994) *Mobilität in der Schweiz. Grundlagenbericht.* Grundlagen zum GVF-Bericht 1/94. Bern: Eidgenössischen Drucksachen-und Materialzentrale.

31   See Truffer and Dürrenberger, *op cit.*

32   This case study is based on several sources, including Hoogma, R., De la Bruhèze, A. and Schot, J. (1995) *Towards Cleaner Cars and Transport, Country Study Germany.* Enschede: University of Twente; DAUG (1996) *Erprobung von Elektrofahrzeugen der neuesten Generation auf der Insel Rügen und Energieversorgung durch Solarenergie und Stromtankstellen. Abschlussbericht.* CD-ROM. Braunschweig/Zirkow: Deutsche Auto-mobilgesellschaft mbH; Schröder, G. (1996) Gegenwärtiger Stand und Vorläufiger Ergebnisse des Flottenversuchs mit Elektrofahrzeugen auf der Insel Rügen, in *Stromdiskussion – Die Zukunft des Elektroautos – Dokumente und Kommentare zur energiewirtschaftlichen und energiepolitischen Diskussion.* Frankfurt/M: IZE Informationsstelle der Elektrizitätswirtschaft e.V., pp. 27–34; Muders, H. (1997) Was kann man daraus lernen? – Schlussbericht Grossversuch Rügen. *MobilE*, No. 2, p. 7; and Prätorius, G. and Lehrach, K-H. (1998) Operation of Electric Road Vehicles in Germany. Investigation of selected examples. A case study for the project 'Strategic Niche Management as a Tool for Transition to a Sustainable Transportation System'. Braunschweig: Reson.

33   Although this mandate would not apply to the German producers before the year 2003, it

did send a message that in the future electric vehicles should be in any car manufacturer's product range.

34 See Schröder, *op. cit.*

35 DAUG, *op. cit.*

36 DAUG, *op cit.*

37 Thomas Albiez, quoted in Escher, B. (1996) Kurzschluss für das Thüringer Elektroauto 'Hotzenblitz. *Süddeutsche Zeitung*, 24 July, p. 26.

38 Ruschmeyer, Th. (1995) Bundesverband Solarmobil – Mobilitäts- und Verkehrskonzept Die Neue Mobilität: 'Solar-Elektro-Leicht-Fahrzeuge', in Proceedings alternativ MOBIL '95, Karlsruhe, 19–22 January, pp. 153–161.

39 Naunin, D. (1995) Germany's EVs need political muscle, in *Electric & Hybrid Vehicle Technology '95*. Dorking: UK & International Press, pp. 261–265.

40 Statement of Rügen project manager, interviewed by Hoogma, De la Bruhèze and Schot, *op. cit.*

41 According to an electric utilities' official in the electric vehicles magazine *MobilE* (March/April 1997), another reason why the car manufacturers chose conversion-design vehicles was that the Federal Research and Technology Minister wanted to start the project in the wake of the German unification. This gave them little time to develop more energy-efficient cars. See Muders, *op. cit.*

42 Rügen Island: Real Science or Public Relations. Electric Vehicle Progress, **16**(9), p. 3.

43 Muders, *op. cit.*

44 Muders, *op. cit.*

45 Schröder, *op. cit.*

46 Informationszentrale der Elektrizitätswirtschaft. (1995) *Stromthemen*, No. 5, May.

47 Grossprojekt über Alltagstauglichkeit von E-Fahrzeugen in Deutschland im Gespräch. (1996/1997) *MobilE*, No. 6, December/January, p. 5.

48 See Legat, W. (1998) Die Zeit arbeitet für das E-Auto. *MobilE*, No. 1, pp. 22–25.

49 Naunin, D. (1998) EV market in Germany: activities in the industry and their political support, in AVERE, *EVT 95, A WEVA Conference for Electric Vehicle Research, Development and Operation*, Conference Proceedings. November 13–15, Paris (two volumes).

50 Friedrich, J., Friedlmeier, G., Panik, F. and Weiss, W. (1999) NECAR 4 – The first Zero-Emission Vehicle with acceptable range, in WEVA, Proceedings of the 16th International Electric Vehicle Symposium, Beijing, October 12–16. World Electric Vehicle Association (CD-ROM).

51 Statement of an official at battery producer Sonnenschein in 1995. Interviewed by Hoogma, De la Bruhèze and Schot, *op. cit.*

52 This case study is mainly based on the report by Schwartz, B. and Maruo, K. (1998) An Outsider Initiative in the Emerging EV Market – the PIVCO Adventures in Norway and California. A case study for the project 'Strategic Niche Management, as a Tool for Transition to a Sustainable Transportation System', University of Göteborg.

53 The government decided in 1991 that this institution should manage a fund, controlled by the Royal Ministry of Transport and Communications, for supporting new transportation technologies that reduce environmental problems. The fund was to get 10 million NOK per year to support various projects. See Schwartz and Maruo, *op. cit.*, p. 33.

54 Quoted in Knie, A. *et al.* (1997) Consumer User Patterns of Electric Vehicles. Research funded in part by the European Commission, JOULE III. Berlin: WZB, p. 77.

55 Figenbaum, E. (1998) TH!NK – a unique city car, in *Electric & Hybrid Vehicle Technology* '98. Dorking: UK & International Press, pp. 47–51.

56 Gjoen, H. and Buland, T. (1996) *Energiteknologiske dilemmaer – Utvikling av gassbuss og elektrisk bil i Norge*, SINTEF IFIM, February, cited in Schwartz and Maruo, *op. cit.*, p. 22.

57 By contrast, in the 1970s project the partners had tried to develop and manufacture every part of the car, which consumed vast resources. There was no competent industrial network to draw upon at the time.

58 PIVCO considered locating the first assembly plant in either Norway or Sweden, but it was important for Norwegian politicians that the City Bees were produced in Norway. Financial support from the Norwegian government made the difference, so that the plant, where eventually 150 persons should be employed, was built in a small town near Oslo. (Schwartz and Maruo, *op. cit.*, pp. 62–63.)

59 This and the next section are mainly based on Schwartz and Maruo, *op. cit.*, pp. 38–51.

60 Nerenberg, V. and Bernard III, M.J. (1995) Station Cars: Personal Mobility at Less Dollar and Social Cost.

61 The Calstart CEO had come into contact with PIVCO during the 1994 Winter Olympics in Norway.

62 This leasing fee covered the vehicle, insurance, and maintenance services, including 24-hour road service. BART concluded contracts with several companies whose employees used the City Bee as station cars at a monthly fee of $100 per vehicle.

63 BART, Hertz to Launch Station-Car Rental Service. *Calstart News Notes*, 27 September, 1999.

64 Gjoen and Bruland, *op. cit.*

65 Information from http://www.avere.org/norstart/

66 Sperling, Shaheen and Wagner, *op. cit.*

67 Since 1991, the shareholders invested $17.5 million in PIVCO, and the Norwegian government contributed about $12.5 million in grants and $5 million in loans. This information comes from Per Lilleng, PIVCO's former president and CEO, in *Automotive News Europe*, 7 December 1998.

68 Ford's EV programme encountered a major setback in 1994 with the fires of NaS batteries, until then the battery of choice for Ford; see Schnayerson, *op. cit.*, p. 84. The company did not yet have a vehicle to comply with the Memorandum of Agreement, but does now. The TH!NK is not Ford's only stake: the company also has an electric pick-up truck in its line-up and is developing hybrid and fuel cell cars.

69 Undheim, T.A. (1999) TH!NK – the Vision of an Electric Vehicle. Paper for the InTePol Project, funded by the European Commission, DG XII, 23.

70 PIVCO lands order for 600-700 'Th!nk' EVs. *Calstart News Notes*, 19 July, 1999.

71 Hertz to distribute, service PIVCO 'Th!nk' EVs. *Calstart News Notes*, 15 June, 1999; PIVCO EV is now 'Th!nk City', production begins. *Calstart News Notes*, 12 November, 1999.

72 This section is mainly based on Simon, B. and Hoogma, R. (1998) The La Rochelle Experiment with Electric Vehicles. A case study for the project 'Strategic Niche Management as a Tool for Transition to a Sustainable Transportation System'. Maastricht: MERIT.

73 Nicolon, A. (1984) *Le véhicule électrique: Mythe ou réalité*. Paris: Ed. de la maison des sciences de l'homme.

74 Bassac, C. (1988) Experimental Uses of Electric Vehicles In France. Proceedings 9th International Electric Vehicle Symposium, Toronto, November.

75 Duranton, S. (1993) *Le véhicule électrique: une dynamique de transformation de l'entreprise*. Paris: École des Mines de Paris.

76 PSA developed two EV models because of the politics of balance between Peugeot and Citroën; these companies are competitors despite their tight organizational relation. The 106 and AX have mostly identical components, the main difference being the body design.

77 The users were 21 private individuals, including housewives, 8 professionals, 19 companies and administrations.

78 86% of users had second cars; 64% had low-range segment cars, of which only 40% were bought new.

79 This section is largely based on PSA, EDF, Municipalité de La Rochelle (1996) *Opération 50 véhicules électriques à La Rochelle – Bilan final de l'opération – Retour d'expérience*. Unpublished internal document.

80 Schwegler, *op. cit.*

81 Robinson, T. Developing niches, in *Electric & Hybrid Vehicle Technology '98*. Dorking: UK & International Press, pp. 28–31.

82   Lane, B. (1998) Promoting Electric Vehicles in the United Kingdom: a study of the Coventry Electric Vehicles Project. A case study for the project 'Strategic Niche Management as a Tool for Transition to a Sustainable Transportation System'. The Open University, Milton Keynes.

83   Blum, W. (1994) Editorial. *MobilE*, No. 3, pp. 2, 6.

85   Crosse, J. (1999) Citroën leads charge for battery-powered vehicles. *FT Automotive World*, July/August, p. 25.

85   This description is mainly based on Harms, S. and Truffer, B. (1998) Stimulating the Market for Lightweight Electric Vehicles: the Experience of the Swiss Mendrisio Project. A case study for the project 'Strategic Niche Management as a Tool for Transition to a Sustainable Transportation System'. EAWAG, Dübendorf.

86   Surveys among 400 LEV-users and 300 non-users and 50 qualitative interviews were conducted by Harms, S. and Truffer, B. (1996) Consumer Use Patterns of Electric Vehicles. Country Report Switzerland. Report for the Commission of the European Communities, DG XII, 1996. This section is based on their work, especially pp. 82–108. See also Knie, A., Berthold, O., Harms, S. and Truffer, B. (1999) *Die Neuerfindung urbaner Automobilität. Elektroautos und ihr Gebrauch in den USA und in Europa*. Berlin: Edition Sigma.

87   Harms and Truffer (1998) *op. cit.*, p. 46.

88   These groups may be contrasted with the public perception of LEV-users, where the image of 'amateurs' was dominant, followed closely by 'greens', at some distance women and mothers, and at larger distance pensioners and yuppies. See Harms and Truffer (1996) *op. cit.* and Truffer, B., Harms, S. and Wächter, M. (2000) Regional experiments and changing consumer behaviour: the emergence of integrated mobility forms, in Cowan, R. and Hulten, S. (eds.) *Electric Vehicles. Socio-economic prospects and technological challenges*. Aldershot: Ashgate, pp. 173–204.

89   Schwegler, U., Van Orsouw, M. and Wyss, W. (1994) *Grossversuch mit Leicht-Elektromobile (LEM). Vorstudie*. Bern: Studie im Auftrag des Bundesamts für Energiewirtschaft.

90   Muntwyler, U. (1997) The marketing concept in the Swiss fleet test in Mendrisio as an example of introducing EVs in the marketplace, in 5th meeting on 'Lightweight Electric Vehicles in Everyday Use', 18 April, pp. 161–167.

91   Schwegler, Van Orsouw and Wyss, *op. cit.*; also Studienreihe *Grossversuch Nr. 18, Grossversuch mit Leicht-Elektromobilen (LEM) in Mendrisio. 1. Zwischenbericht*. (Bern: Studie im Auftrag des Bundesamts für Energiewirtschaft, 1997).

92   Schwegler, Van Orsouw and Wyss, *op. cit.*

93   Harms and Truffer (1998) *op. cit.*, p. 32.

94   Fahrzeugsubventionen in der ersten Hälfte 1997. *LEM-News* (1997) Nos 17/18, p. 28.

95   Begleituntersuchung: Zwischenbericht der 1. Versuchsphase veröffentlicht. *LEM-News* (1997) Nos. 17/18, pp. 31–32.

96   The energy consumption limit for 100% subsidy was 18 kWh/100 km in 1997 and 16 kWh/100 km in 1998 for two-seaters; for four-seaters 22.5 kWh/100 km in 1997 and 20 kWh/100 km in 1998 (measured according to the Biel test). See: Regulations for test vehicles of the Large Scale Test with LEVs in Mendrisio. *LEV-News* (1997) Nos.19/20, p. 21.

97   Blum, W. and Wegmann, S. (2001) Der Beitrag von Elektrofahrzeugen zur nachhaltigen Mobilität.: *SEV/VSE Bulletin*, No. 8, pp. 25–27.

98   See Gran Consiglio della Repubblica e Cantone Ticino 2000. Messagio sul Estensione del progetto VEL a tutto il cantone Ticino. Rapporto numero 5020, 27 Giugno 2000. Dipartimento del Territorio, Bellinzona.

99   Piffaretti, M. (1995) Mendrisio – a true market demonstration, in *Electric & Hybrid Vehicle Technology '95*. Dorking: UK & International Press, pp. 61–63.

100 Simon, M. (1997) Bedeutung und Wirkung von Fördermassnahmen, in *Leicht-Elektromobile im Alltag; Elektromobil – Mikrowellenherd im Mobilitätssektor*, Tagungsunterlagen Band 11, Zürich, 17/18 April 1997, pp. 248–252.

101 Studienreihe Grossversuch Nr.15, *Bekanntheit und Beurteilung von LEM, des Grossversuchs, der Fördermassnahmen sowie Ergebnisse zu den Partnergemeinden* (Bern: Studie im Auftrag des Bundesamts für Energiewirtschaft, 1996).

102 Pulfer, M. (1997) Zwischenstand und Auswertung der Fahrzeugzulassungen Mendrisio und den Partnergemeinden, in: *Leicht-Elektromobile im Alltag; Elektromobil – Mikrowellenherd im Mobilitätssektor*, Tagungsunterlagen Band 11, Zürich, 17/18 April.

103 Meier, E. (1997) Grossversuch mit Leicht-Elektromobilen (LEM) in Mendrisio: Fahrleistungen und Energieverbrauch', in: *Leicht-Elektromobile im Alltag; Elektromobil – Mikrowellenherd im Mobilitätssektor*, Tagungsunterlagen Band 11, Zürich, 17/18 April 1997; only LEVs that count as cars are taken into account.

104 The local garages were willing to support the experiment as their car sales decreased by 20% in the preceding years. However, they were critical of the technical state of LEVs. Milton Binaghi, garage keeper in Mendrisio, states: *'In order to make LEVs more interesting, the distance range must be enhanced. The main problem, however, is the price of the battery which is, in my experience, the most important obstacle for purchasing a LEV.'* See Studienreihe Grossversuch Nr. 18, *op. cit.*

105 Hüsler, G. (1996) *Bericht über die durchgeführten qualitativen Interviews mit privaten Käufern von elektrischen Fahrzeugen im Grossversuch in Mendrisio.* Research report of the large-scale test in Mendrisio.

106 Hackney, J. *et al.* (1999) *Interviews mit 15 LEM-Besitzern im Rahmen einer Untersuchung ihrer Fahrleistung. Aktionsprogramm Energie 2000 Leichtmobile*, Bern.

107 Fahrzeugsubventionen in der ersten Hälfte 1997. *LEM-News* (1997) Nos. 17/18, p. 28.

108 Studienreihe Grossversuch Nr. 18, *op. cit.*

109 Editorial. *LEV-News* (1998) No. 23, p. 4.

110 Neupert, H. (1997) Anziehender Markt. *MobilE*, No. 2, pp. 18–19.

111 Students developed the Twike in the early years of the Tour de Sol (1985), but it took several years to prepare production, due among others to difficulties in raising the necessary capital. Eventually they followed a strategy of asking interested customers to deposit a substantial sum of money to finance the production of the LEVs. The first series of 200 Twikes was sold within a short time in 1997, and production of a second series of 400 vehicles started in autumn of the same year, again by order. Some investment firms formed a joint venture with the old company. For marketing and after-sales services the company set up a network of 30 local 'competence centres' (dealerships). See: TWIKE: Production of the next series take a run-up. *LEV-News* (1997) Nos. 19/20, p. 4.

112 W. Blum,W. (1996) Twike III. *MobilE*, No. 3, pp. 8–11.

113 TWIKE + Mobility = Mobilität. *LEM*-News (1998) No. 22, p. 7.

114 TWIKE-Challenge: 2 Months of Adventure and Much Response. *LEV-News* (1998) No. 23, pp. 4–5.

115 This includes sites where a fee is due, such as restaurants or parking houses, but the majority of sites are still offered for free by private individuals. Some tourist centres, warehouses, municipalities and electricity companies also offer parking spaces for charging. See www.e-mobile.ch, and also Zeller, P. (1997) Vom Park & Charge zum Charge & Drink Konzept des Aufbaus einer Elektromobil-Infrastruktur, in *Leicht-Elektromobile im Alltag; Elektromobil – Mikrowellenherd im Mobilitätssektor*, Tagungsunterlagen Band 11, Zürich, 17/18 April 1997, pp. 222–225.

116 SULEVs have emissions less than one-fifth the level of 'LEV tier 1' standards. Automakers receive a 0.2 ZEV allowance for every car sold that meets the SULEV standard. Hybrid cars that meet the SULEV requirements earn the producer 0.3 ZEV allowances because of the all-electric capability. Furthermore, carmakers are allowed to decide which mix of vehicles to use to meet the 10% ZEV requirement for the 2003 and subsequent model years. The exception is that large volume manufacturers have to meet at least 40% of the requirement using vehicles receiving a full ZEV allowance (battery electric vehicles or fuel cell cars). See: California Air Resources Board (1998) Public Hearing to Consider the 'LEV II' and 'CAP 2000' Amendments to the California Exhaust and Evaporative Emission Standards and Test

Procedures for Passenger Cars, Light-duty Trucks and Medium-duty Vehicles, and to the Evaporative Emission Requirements for Heavy-duty Vehicles, November.

117 Fuel cells are much more efficient than internal combustion engines and can thus substantially reduce CO_2 emissions (in the order of 30%), while other regulated emissions are absent. Fuel cells have no moving parts, do not vibrate, make no noise, are virtually maintenance free, and do not have the drawbacks of battery electric vehicles such as short driving range, heavy weight, short battery life and power plant emissions. Fuel cells are suitable for mass production and may then be as cheap as today's gasoline engines. Different types of fuel may be used as a source of hydrogen, requiring different solutions to questions of supply, storage and conditions of use. There are still several technical issues to solve before these potentials can be realized, however, including fuel storage and price of fuel cells.

118 Hoogma (1999) *op. cit.*

Chapter 5

Experiments in reconfiguring mobility

The experiments of the early 1990s aimed at electrifying mobility failed to lead to a regime-shift as some actors, for example, in Mendrisio, had hoped. It appears that the actors who upheld the existing regime had absorbed the ideas generated and the expertise gained during the decade of experimentation with battery-electric vehicles and had begun using these to convert the gasoline car into a hybrid and/or fuel cell vehicle. By the start of the year 2000, these two vehicle technologies were expected to gain a substantial market share in the near future. This development could be interpreted as a regime-optimization scenario, since these cars were seen only as replacements for existing cars, and, as such, would not contribute to changing Western mobility patterns.

At the same time, technological niches for battery-powered vehicles have remained alive, and several actors have invested in them, so their chances of success may again improve in the future. We can ask, 'What if the PSA and Mendrisio experiments had resulted in more EV sales?' or 'What if the German government had responded favourably to the German manufacturers' reluctant 1994 initiative to create a push for electric vehicles?'

Such questions demonstrate the crucial role played by fundamental historical contingencies that are part of any regime-shift. But the shift did not happen. For one thing, niche developments at the micro level failed to mesh with, or gain positive feedback from, developments at the regime and socio-technical landscape levels. Another reason is that optimism about the chances of solving environmental problems within the dominant regime remained high throughout the 1990s, so automobile manufacturers considered improving the existing regime as their main option. In addition, the urgency of solving environmental problems such as, for example, the greenhouse effect, did not increase much during the 1990s.

At the level of niches and experiments, it is clear that most actors did not aim for radical change either. They failed to exploit windows of opportunity that opened in the PSA experiment and in Switzerland. Both producers and users remained locked into the idea that the car was the generic means of individual transport. They assumed that this one single technical artefact should be able to cover the lifetime mobility needs of most individuals and households. This assumption led to car designs dictated by peak-use requirements: cars with high top speeds, sportscar-level acceleration, large fuel capacities to extend the distance between refuelling stops during long trips, and with enough room to transport several passengers and large quantities of goods. Knie and Hård used the term 'race-and-travel limousine' (*Renn-Reise-Limousine*) to label such vehicles, which have been the paradigm for automobile engineers since the early twentieth century.[1]

Consequently, today's gasoline cars are strongly overpowered given the needs of most drivers and passengers. For example, internal combustion cars can usually travel 500 km or more on one tank of fuel. Many users consider this a minimum requirement, although 95% of trips in a country like Switzerland are shorter than 50 km. Another example is the recent trend towards off-road vehicles and MPVs, both of which have experienced striking gains in market share in the United States and Europe. Minivans can carry eight people and their luggage, although in Europe the average number of passengers per car per trip is 1.2 people. And although off-road vehicles are designed for use on rough or sandy roads outside built-up areas, most traffic flow takes place on well-maintained roadways in urban areas. Thus, today's cars are highly sophisticated material- and energy-intensive artefacts that are most often used far below their potential operating capacity. This in turn creates a major problem of idle capacity, with further implications for energy consumption and for vehicles' material intensity.

One promising option for working toward a regime-shift is to use information technologies to combine new and existing transport modalities more efficiently and thus to offer mixed forms of public and private means of transport. This could result in reconfigurations of mobility because both the producers and the users would have to change their routines quite radically. However, like the electrification option, such a provision could also lead to regime optimization if new technologies are used only to improve the private and public transport regimes separately. One important area to examine, therefore, is the findings from recent experiments aimed at building niches with the potential to contribute to a regime-shift.

This Chapter discusses four such experiments: an inner-city bicycle-loan scheme in the UK; a flexible system of community transport (CT) for disabled and elderly people in London; a bottom-up initiative for organized car-sharing in Switzerland; and an individualized public transport service based on EVs in France. Before focusing on these experiments, we will place the relationship between private and public transport in historical perspective and suggest that the introduction of new information technologies presents a window of opportunity to bridge the divide between private and public transport.

Competition between gasoline cars and public transportation

Chapter 4 sketched the growing dominance of the gasoline car early in the twentieth century by focusing on drive trains and vehicle concepts. As the century progressed, gasoline cars became more differentiated with regard to both comfort and status. In the 1920s, GM introduced a whole range of car models from low-priced worker models to high-priced executive models. This product philosophy proved highly successful and was soon copied by other car manufacturers. Upward social mobility could now be demonstrated publicly by the kind of car a person drove; employees and management could be recognized on the street even before they left their vehicles.

By the 1950s and 1960s, mass motorization had become a societal project that promised unrestricted participation in the new consumer economy. Although work in mass-production factories was tedious and repetitive, workers saw the rewards of their labour in their homes, which became filled with modern conveniences and appliances such as televisions, radios, washing machines, etc. The massive expansion of single-family dwellings and the unprecedented pace of suburbanization brought about major changes in living conditions and lifestyle. The car became the precondition of a new lifestyle, where living and workplace moved farther and farther apart. This development brought increasing traffic to city centres but undermined living conditions for those still living there. The consequence was increasing urban flight, reinforced by 'urban renewal' projects undertaken by local and sometimes state and federal governments to clear areas needed to meet the growing need for space for new city highways. The results of these mutually reinforcing developments, which have led to more cars and more decentralized urban structures, can be seen most visibly in Los Angeles.[2]

These developments had negative consequences for both city centres and for population segments that did not want or could not afford to leave the inner city. Local citizen groups organized active but often futile protests against urban-renewal projects formulated in the early 1960s. In the 1970s, the oil crisis and a growing awareness of the 'limits to growth' gave new impetus to critics of the gasoline car in the major industrialized countries. Suddenly the car stopped being only a symbol of wealth, progress, and upward mobility; rather, it was also accused of killing people, destroying ecosystems, exhausting natural resources, and promoting selfish and socially irresponsible lifestyles.

Although these criticisms were necessary precursors to reconceptualization of the role of the automobile in society, they failed to halt, much less reverse the spread of the automobile. Instead, reform proceeded incrementally. Policymakers pressured car manufacturers to make cars cleaner and more efficient. Overall, these attempts have been quite successful and have increased the pace of innovation in vehicle and propulsion system design. But the resulting gains in efficiency have been used to increase size and add new features, not to improve system efficiency. Moreover, more drivers have begun using more cars for more trips over longer distances. Having fulfilled the dream of one car for every household, Western societies are moving towards a one car per person ratio. Sports and leisure vehicles are gaining growing shares of the market, clear proof of the increasing popularity of leisure activity related travel. Finally, the small, city car segment has only recently begun to gain a share of the market.

In most nations, the rise and reinforcement of automobile-based transportation has been linked with a decline in public transport. Trams and trains were the first mass transport means – long before cars came on the scene. These modes involved investment in underground systems and city trains. With the increase in car traffic, however, investment in public transport infrastructure was reduced in favour of road construction. Public transport lost its appeal. From the 1970s, however, public transport has regained some of its importance, in part as a reaction to environmental issues and to the inconveniences of urban traffic regulation. This has resulted in reserved lanes for buses and increased numbers and comfort of urban and inter-city trains. The main aim, however, has not been to create a new regime that can address the shortcomings of today's public transport but rather to safeguard public transport's market share *vis à vis* the automobile.

From the consumer's point of view, public transport at the start of the twenty-first century suffers from numerous and definite shortcomings, most of them due to its intrinsically collective character. Consumers perceive the travel times of public buses, trams, and trains as longer than those of private cars; only the underground is perceived as a fast means of transport. Infrequent users view public transport as unreliable, uncomfortable, excessively costly, and complicated to use. Furthermore, the limited flexibility of public transport with regard to both scheduling and routeing is seen as a major drawback. And finally, public transport is perceived as lacking in privacy and potentially dangerous, and is also associated with low social status.

In the final decades of the twentieth century, most actors realized that public collective transport had little potential for replacing all the trips made by car. Rather, its advantages lay in dealing with traffic flows that were highly concentrated and where there could be economies of scale. However, in locations characterized by highly diversified mobility patterns, public transport can lose even its environmental advantages over the private car if buses and trains are travelling the countryside half-empty or worse and their capacity is going unused. In short, given the historical trend of increasingly individualized lifestyle choices and the growing dissociation between living and working places, public transport is not in a good position to compete with the private car.

Summarizing the historical development of and relationship between public and private transport, it is obvious that the divide between these two regimes has grown ever wider. Public collective transport cannot compete with the private car for the reasons mentioned above. Thus, the prevailing patterns of development in forms of mobility, land-use practices, and innovations in the car industry have resulted in an entrenched 'division of labour' between public and private transport.

At the user level, powerful processes of habit formation can be observed, and perceptions of performance, cost, and comfort are strongly skewed in favour of the private car. As a consequence, users increasingly rationalize their existing mobility patterns as if there existed an objective need to own and use a car in the currently defined and accepted ways. Public transport, the dominant option in passenger transport a century ago, has become a secondary mobility option. Today it is limited to satisfying specific travel needs such as short-distance intra-urban or inter-city transportation, or to serving segments of the population – children, the elderly, the disabled, and the poor – who cannot afford to own or cannot drive a car.

We expect that public transport has some potential for regime upgrading but will never replace a large proportion of today's car traffic. This appears all the more true if current socio-economic trends, such as the continuing individualization of lifestyles and the spatial separation of living and work places, are projected into the future. These factors will lead towards increasing reliance upon private transport and tend to discourage a revival of public means of transport.

Mixing public and private transport – a new trend?

With the emergence of new information technologies in the 1990s, many different actors have identified new options for upgrading public transportation. For example, public transport passengers must be better informed about the current state of the system. They want to know how they can get from point A to point B in the fastest, most efficient way. The convenience and reliability of timely arrival would be enhanced if users had access to immediate information about adjustment of the travel schedule if one of the transportation links is unexpectedly delayed or out of service. Furthermore, public transport users are known to prefer a single, simple system for determining the cost of trips and of actually paying for tickets.

New information technologies can meet these needs, and in fact quite a number of systems have already been developed or are under construction. These innovations will heighten the appeal of public transport. We can arrive at similar conclusions when we look at the potential contribution that information technology can make to private transport. For example, electronic and computerized routeing and mapping devices can inform the driver of optimal routes and eliminate lengthy on-the-road searches. In addition, such information and traffic-guiding systems may optimize use of existing transportation infrastructures and postpone the need for building new roads. Modern car systems can also interact with drivers, analyse the technical state of the mechanical system, and call for technical assistance if major problems appear.

Information technologies are thus likely to reinforce existing mobility patterns and lead to regime optimization of both public and private transport. Eventually, public transport may become more attractive and more convenient to use, but so will private transport. We may expect an increase in overall demand for transport services but only minor changes and reconfiguration in mobility patterns.

This is not the entire story, however. Building on the potential and promise of information technologies, various actors have also been working to find ways to bridge the public/private transportation divide. They realize that information technology can be used both to counteract the inherent weaknesses of public transport and to highlight the inconveniences of private transport.

One promising approach is to work on the alleged cost efficiency of private cars compared to public transport. In practice, consumers generally underestimate the cost of owning and operating a private car and overestimate the costs of public transport. This perception is a result of the particular ways costs for cars are paid for versus the costs of public transport. Most car-based costs arise either once every five to ten years when a new vehicle is purchased, or, like insurance, taxes, fees, certain maintenance costs, etc., are due on an annual or fixed basis. Fuel and parking are the only costs felt more or less on a per-trip basis. In general, therefore, the relatively high fixed costs of vehicle ownership encourage vehicle use, as each additional mile driven decreases the average total costs incurred. But because the fixed costs are just that – fixed – car owners underestimate the total burden of car ownership on the household budget and are inclined to use it as often as possible.

Public transport, on the other hand, has a linear cost-mileage relationship, and the cost of a ticket represents the full cost on a per-trip basis. Thus, on a one-to-one basis public transport appears to be much more expensive than a private car. This is all the more true after the members of a household decide to purchase a car. The household then gets locked into the rationality of car use, which is almost impossible to change as long as the car is running and no further changes take place in the living situation.

If a comparison between different means of transport is carried out on a total cost basis, however, the advantages of owning a private car are much less obvious. In addition to direct fixed and variable costs, car owners face additional inputs of time and energy if they are to receive adequate transport services from their vehicle. Some of the other 'costs' include maintenance such as washing the car, unscheduled repairs, and so forth. There is also the sheer effort of driving, an activity that many drivers rationalize as enjoyable even though driving in normal traffic conditions in modern conurbations can be quite stressful. The time spent driving is also time that is unavailable for other activities like reading or relaxing.

Some actors realize that alternative forms of mobility, such as public transport, could capitalize on these hidden costs, but an effective argument would first have to overcome average car users' taken-for-granted perceptions of comfort and costs. If users could be led to question the fundamental preconditions of their preferences, they might then come to see alternative modes of transport as superior to owning a private car. But such a change would be difficult to achieve through conventional means such as marketing and consumer information. Consumers would have to be able to experience the new means of transport directly, and new definitions of comfort and lifestyle would have to be communicated in the actual use contexts.

One major disadvantage shared by alternatives to the private car is the high transaction costs for getting access to the respective means of transport. Having an all-purpose vehicle standing in front of the house ready to be used for whatever need arises is a seductive option that also carries important symbolic value – the freedom to go wherever one wants whenever one wishes.

Public transport, on the other hand, with its limited number of stops and its fixed and often ill coordinated time schedules, is perceived as the antithesis of that freedom. Thus, new options must be positioned as flexible and customer-oriented and must at the same time emphasize the specific advantages of different modes of transport, for example, that public transport offers gains in leisure and working time because users need not drive in stressful traffic conditions. The classic (i.e., fixed route) form of public transport has inherent systemic limitations that prevent it from becoming as flexible and user-tailored as the private car. Nevertheless, actors have started to realize that there is considerable potential for innovations that might bridge public and private transport.

Another issue is the need to increase the public's ability to deal with a system that integrates multiple modes of transport, because this could in the medium term support a shift away from predominantly car-based mobility patterns. Today's car owners have a considerable degree of know-how for dealing with complex use conditions. Imagine a driver who must be at a meeting but becomes stranded in city traffic during rush hour when his car breaks down. Most people would accept such an event as an inevitable and accepted, if unwelcome, part of life in modern society, but the same people would consider a similar situation as a sheer horror were it to happen in a public bus, tram, or train. Thus, car drivers may shy away from using public transport because they are concerned that unpredictable problems may pop up, and that they would not know how to deal with them. It stands to reason that if drivers made more frequent use of other means of transport, their ability to deal with the public transport system would increase, and their willingness to accept difficult situations while driving their private car might decrease. In short, the learning and unlearning of specific capabilities might result in substantial reversals and re-evaluations of existing perceptions of the costs and comforts of driving.

Various actors have come to realize that, taking these user-related preconditions together, there is room for new learning processes that could persuade a substantial proportion of the car-driving community to leave their cars at home, or even replace them with other means of transport. Whether this happens will depend, among other factors, on the development of new technological forms that bridge the gap between the two transport regimes as they exist now, and also on the development of new opportunities for users to experience them.

We now turn to an examination of the four experiments mentioned above that were set up to test some of these new hypotheses.

Experimental strategies

Three major approaches are available to bridge the gap between public and private transport. The first includes initiatives for making public transport more flexible and better adapted to consumers' needs, but, as mentioned above, public transport suffers from structural weaknesses such as fixed time schedules and collection points. One domain where considerable innovative activity has taken place is in services such as 'dial-a-bus' and collective taxi experiments, for example the Dutch train-taxi. Information technology plays an important role here because of its great potential for optimizing route planning and vehicle tracking for both service providers and users. However, users must first learn to use this kind of system and to calculate the real costs of this mobility option if they are to become convinced that it is an effective, appealing, and affordable alternative.

The second category focuses on the collective use of private means of transport. Examples include car-sharing experiments among several households in a neighbourhood or within a family, ride-sharing (i.e., car-pooling), an internet-based brokerage arrangement for sharing long-distance trips (Germany's *Mitfahrgelegenheit*), and voluntary schemes for transporting disabled or elderly people. These schemes could be greatly enhanced by information systems that would offer users improved access to vehicles, tracking of vehicle locations, and assistance in the event of trouble or accidents. These strategies rely strongly on the ability of users to adapt. They need to accept that their formerly private space becomes semi-public, as in the case of car-sharing and car-pooling. Or they have to learn to select appropriate options for different transport needs, for example, reliability, timely access and appropriate transaction costs in the case of car-sharing or long-distance trip-sharing.

The third approach focuses on actively pressuring drivers out of their cars and onto public means of transport. A number of inhibiting regulations can be named, such as discouraging certain segments of drivers from entering specific areas (for example, bans on inner-city travel for commuters, odd-even licence-plate access days, access only for specific vehicles such as EVs, small city cars, or cars with a minimum number of passengers, etc.). Such schemes require the existence of collection points where those who are not entitled to continue with their cars can change to public transport. Information technology is likely to improve the smoothness and appeal of such systems. In the case of access restrictions, information technology will make monitoring and prosecuting violators easier and thus improve the effectiveness of the measures. However, if the majority of drivers remain committed to automobile-based mobility patterns, these initiatives are unlikely to find widespread acceptance, and drivers will be highly motivated to circumvent the regulations.[3]

Several recent experiments have focused on trying to close the gap by stressing the freedom offered by new information technologies and thereby encouraging users to develop new mobility patterns. We will discuss in detail four experiments that represent typical strategies for bridging the public-private transport gap. Figure 5.1 locates these four experiments in the public-

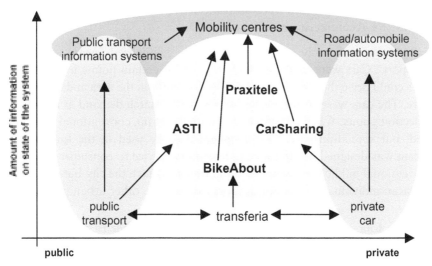

Figure 5.1 Approaches to bridging the gap between public and private means of transport that are enhanced by the increasing use of information technology and may lead to the emergence of new mobility patterns

private gap in transport space. All four could lead to the development of a new kind of actor and service that is presently not available: a mobility centre.

These four experiments are:

1 *Bicycle pooling*. A rudimentary transfer system between private cars and a bicycle pool was created to persuade car drivers to use bicycles for inner-city trips. This experiment offers useful lessons for other attempts to enhance the appeal of public transport using information technology and to induce drivers to change their preferred means of transport.

2 *ASTI*. This experiment aimed at making public transport more flexible by creating a 'semi-public' customer-tailored means of transport. It took place in a community transport system in the UK that provided access for disabled and elderly people. This experiment can be looked at as a test of a number of initiatives that hope to enhance public transport by providing door-to-door services.

3 *Organized car-sharing*. This experiment aimed at dissociating ownership from the use of cars. Members of the car-sharing organization could get access to a car whenever they needed it by using a centralized reservation system. The ownership of the vehicle was in the hands of the organization, usually a user-owned co-operative, although commercial organizations also exist. By separating ownership and use, the inherent tendency to use a car because of the high up-front investment cost is eliminated. Cars in such a system may be used much more efficiently. This particular experiment analyses the emergence of car-sharing organizations in Europe, and Switzerland in particular. These initiatives have been quite successful in recent years and hold out a strong promise for reconfiguring mobility patterns.

4 *Praxitèle*, a French experiment with 'individualized means of public transport'. Cars were made available to users at certain points in a city, and users could drive them to any other collection point in the area and leave them there. The cars were then relocated as needed to match demand at individual collection points. When the system was initially set up, conventional cars were used, but special-purpose EVs are planned to be used in the future. This system was designed to offer a complementary service to consumers who use conventional public transport buses and trains to reach the city but then want to use an individual means of transport for travel to other urban points.

These four experiments illustrate different approaches to closing the gap between public and private transport. As transport-related information systems for both public and private transport develop further, these mixed forms may give rise to a future mobility system that we will call an 'integrated mobility system'. This involves the inclusion of all means of transport into an individually tailored 'package deal' in which users determine for every trip which transport modes or combinations of modes best suit their needs. Such a system would minimize problems of idle capacity, and users' habitual behaviour would move away from a car-dominated system.

Several locations have already established first attempts at integrated mobility centres. However, their smooth functioning and broader diffusion will depend on the development of reliable bridging technologies and new practices. The four experiments focused on developing and testing such practices and technologies; the discussion below follows the basic format introduced in Chapter 4's descriptions of the experiments on electrifying vehicles.

Portsmouth's Bikeabout bicycle-pool scheme: getting people out of their cars

Introduction[4]

The bicycle is probably the ultimate clean-transport technology. An average journey in town by an ordinary car consumes up to 80 times more energy than the same trip by bicycle,[5] and the energy used to propel the bicycle is obtained from non-fossil fuels. The dominant policy approach to promoting bicycle use has involved municipalities' providing physical measures such as cycle ways and cycle lanes. These measures are often connected to information campaigns that, for example, stress the health benefits of cycling. Other policy measures, such as alternative means of obtaining access to bicycles or subsidizing bicycle use, are rare, although such measures are commonplace in campaigns to promote the use of public transport or car-sharing.

Promoting bicycle pools and bicycle-loan schemes offers a route towards extending cycling as a system. In various cities such as La Rochelle, Amsterdam, Zurich, Copenhagen, and Cambridge, city councils have provided fleets of bicycles free or for a nominal charge. The idea was that bicycles could be obtained from and returned to designated 'public bicycle' racks around the city. Such schemes have faced serious operational difficulties, however. In most cases the schemes failed because the bicycles were stolen or vandalized in their open racks. Other problems have included negligent maintenance, undetected defects in the bicycles, and users not being able reliably to obtain a safe and roadworthy bicycle.[6] The lesson learned from these cycling schemes was that public bicycle schemes need a more controlled and protected environment if they are to succeed.

The Bikeabout scheme in Portsmouth, England, sought to address these issues. It emerged from Portsmouth University's Mobility Policy, which was developed in 1992 because the University needed an encompassing strategy for reducing staff and student vehicular traffic in advance of an expected increase in traffic due to the planned expansion of the University to two sites 3 km apart. Part of the policy focused on enhancing the use of bicycles at the University by improving the facilities for parking cycles.[7]

The task of refining and aiding the implementation of the strategy was assigned to a trainee who had recently joined the University. The trainee worked with transport researchers in the Department of Geography to carry out a detailed survey of staff and student travel patterns. The data showed that staff and students made a substantial number of trips between the two sites during the day. The trips were usually short and therefore suitable for cycling, but one in four trips between the two sites was made by car. The idea of starting a cycle-pool scheme for these trips was broached, while, at the same time, Hampshire County Council was looking to co-fund an innovative Portsmouth-based energy-saving project as part of the European Commission's ENTRANCE project.[8]

At the trainee's initiative, Portsmouth City Council requested funding for an innovative cycle-pool scheme that would eliminate theft and vandalism and control the use of the bicycles. The city officials also saw the project as an excellent opportunity to enhance cycling facilities in the city with funding that would not otherwise have been available. A dedicated cycle route was incorporated into the project on the grounds that it would be irresponsible to implement a cycle scheme without considering the users' safety.

The main obstacles to starting the plan were neither technological nor due to any absence of motivation, but rather to institutional factors. After the initial impetus to get the scheme going, the plan was put on hold for 15 months until funding to match the European contribution was secured.[9] Portsmouth University and the City Council also contributed additional funding. The so-called 'Bikeabout' concept was integrated into the larger mobility policy, and the trainee was assigned to develop a plan for implementing the scheme.

Objectives and project organization

The overall scheme was intended to minimize the environmental costs and the energy usage of staff and students at the University. Its aims can be summarized as follows:

1 To encourage greater awareness of alternatives to car use for travel between the two University sites;

2 To make bicycles more accessible to people needing to travel in Portsmouth during the working day;

3 To provide a safer environment in which bicycles could be used.

A 'Bikeabout Steering Group' was formed to guide the progress of the scheme. It included representatives of the two project partners, the city council and the University, as well as the former trainee who now was a freelance consultant. The monitoring and evaluation were contracted to the Transport Research Laboratory, a company that subcontracted the work to the University's Department of Geography.

The first phase was launched in October 1995 with the opening of two depots, at opposite ends of a designated cycle way, stocked with 40 bicycles ready for use and bearing distinctive University identification. In July 1996, this system was replaced with a fully automated one that included 100 bicycles. Within the first year, 500 users had registered for a smartcard that gave them access to the bicycles. A third depot, funded by separate ENTRANCE monies, became operational in December 1997. Bicycles are obtained from the electronically controlled racks simply by entering the smartcard into a console at the depot. The use of the bicycles is free.[10] Every day the detailed transactions from each depot are sent by modem to a central computer to ensure that all the bicycles are accounted for. The smartcard system guarantees that members can be held responsible for the bikes they use.

The distinctive feature of the Bikeabout experiment is that it has applied advanced technology to provide a controlled and protected environment in which to introduce a cycle-pool scheme. Covered depots offer protection from the weather and from theft. Closed-circuit television surveillance has also curtailed vandalism and contributed to security. The smartcard system was chosen over an on-line system because it was feared that the latter might not function at all in the event of communication failures. Another benefit of the smartcard is that it supplies enough information to prevent users from borrowing more than one bicycle, regardless of which depot is used. The stored information also provides a verifiable record of the transaction and bicycle number in case of any disputes.[11]

Users can report defective bicycles to the system console and obtain another. When the system is notified of a defective bicycle, the unit is automatically taken out of service until it is repaired. This is an essential part of the system because it ensures regular maintenance of the bicycles. The system can also issue warnings to the administrator when a depot is becoming full or empty so that bicycles can be redistributed. Users can also register their preferences so they can get the type of bicycle they prefer. Three conventional designs are available, all with battery lights, an easy-to-adjust seat, and a large front carrier.

One significant barrier to cycle use is the danger of traffic, particularly in the centre of a large, busy city like Portsmouth. Both the City and County Councils addressed this issue through their involvement in the experiment. The City Council developed a network of cycle routes throughout the city, and provided a safe cycle route between the two initial Bikeabout depots. This was achieved through several measures, including cycle lanes and two signal-controlled crossings where the route crosses heavily trafficked roads.

Evaluation of project design and management

Managing the project was a relatively easy task because there were few partners and the advanced technology was located in the depots. Organizers felt that one of the most important lessons learned during the Bikeabout project was the need to retain flexibility. Because the scheme was so innovative, they had no relevant examples of previous practice on which to rely. Therefore, most elements of the project had to be designed and tested incrementally. This meant that the whole scheme had a long lead time, and the organization of the scheme was very time-consuming. Furthermore, the start of the project was delayed by the time it took to secure funding. However, the commitment and enthusiasm of the individuals representing the project partners made up for these difficulties.

One effect of the co-operation between the two partners was that, although the University did not have an official responsibility to fund the Bikeabout scheme, any moves to disband it would have harmed the relationship with the City Council and were thus avoided. A good relationship was vital if the University was to continue to expand, as it plans to do by redeveloping a new site near the main campus for relocation of its business school.

Learning

The staff and students of Portsmouth University started off with quite low expectations for the shared bicycle scheme. They generally felt that such schemes did not work because bicycles were poorly maintained, generally unreliable, and often stolen. But the Bikeabout users soon learned that the system was reliable and worked well, and not one bicycle was stolen or vandalized. The experiment confirmed the importance of good maintenance. In some earlier cycle pools, the bicycle design discouraged theft, but in Bikeabout it was considered essential to provide high-quality bicycles if drivers were to be persuaded to use them instead of their cars for trips around the University. After the design and testing of Bikeabout were completed, Dixon-Bate Ltd., the private company that won the contract to provide the electronic cycle stands, was so excited by its further commercial prospects that the company planned to market Bikeabout to other communities.[12]

Although the project partners and users learned that the Bikeabout system was viable, mobility patterns were neither questioned nor changed. Second-order learning failed to take place. The results indicate that the bicycles were used as an alternative to walking, not to driving. Only one-fifth of Bikeabout users had been using their cars for inter-campus trips, and most of them reported that the main attraction of bicycling was to travel in a more environmentally friendly manner or to get more exercise. Interestingly, non-users of the Bikeabout scheme increased their use of bicyles, probably because of the improved cycle paths and facilities. The most common reason reported for not using the scheme was that the respondent already owned a bicycle.

The experiment suggested that when it comes to societal acceptance and political conditions, merely providing a cycle-pool scheme is not enough. It was unrealistic to expect Bikeabout to effect a major change in travel behaviour. To achieve that result, it appears that other complementary measures, particularly those aimed at discouraging car use, are required.[13]

Societal embedding

The project was carried out by a network of participants with the University and City Council as the main actors and with very committed individuals doing the work within these organizations. Their initial motivations can be understood in terms of a convergence of interests arising from the University's attempts to establish a mobility policy and the council's interest in useful projects that could quality for EU funding. The relationship between the City Council and the University was enhanced through their work together on the Bikeabout scheme. In addition, potential users were involved through meetings with representatives of the student union and other interested parties.

Dixon-Bate Ltd. played a significant part in the project. Although not technically a partner in the scheme, the company went far beyond its duties as a supplier and worked with the software supplier to refine the technology used in the scheme. The company has a long tradition of promoting the bicycle as an environmentally friendly transport alternative. Under the terms and conditions of the EU funding agreement, the official partners of the scheme could not themselves profit from the scheme. In addition, neither the University nor the City Council considered commercial development as within their mandate. They thus allowed Dixon-Bate to obtain the complete design blueprint and to market Bikeabout as its product. The company subsequently sold the system to the city of Rotterdam, where several Bikeabout depots have been built. However, there is no obvious institution or network through which dissemination of knowledge about Bikeabout is occurring apart from an informal network around the NGO 'Transport 2000'.

There is only tentative evidence to suggest that Bikeabout improved the image of cycling at the University and contributed to promoting a cycling culture in Portsmouth. Awareness of the bicycle's importance as a mode of transport did seem to increase, and there appears to have been a significant increase in the number of people cycling. But with only 500 registered users after a year, the success of the cycle pool itself was not considered outstanding.

Expectations about the functioning of the cycle pool were initially low due to earlier experiences elsewhere. A survey of staff and students revealed that the cycling facilities, especially bicycle paths, offered by the wider University Transport project were well-received but that the cycle pool would scarcely foster shifting from cars to bicycles for inter-campus travel unless accompanied by disincentives such as parking fees for car use. Bikeabout raised expectations because the smartcard and the surveillance technology solved the security and maintenance problems of earlier schemes, but the number of users was less than expected.

A niche difficult to diffuse

The Bikeabout experiment overcame the main problems of shared bicycle schemes by providing the core requirement of secure on-demand access to working bicycles. On the technical side, it has been a virtually unqualified success. Uncertainty remains, however, about applying this scheme as a 'tool' to solve transport problems. A scheme like Bikeabout, as it was organized, cannot be expected to succeed as an isolated measure. It must be integrated with other transport management measures. Significantly, smartcard systems have the potential of offering a totally different way of obtaining access to mobility. The smartcard technology and operational system developed for Bikeabout could be adapted relatively easily to other types of rental or loan schemes, for example, in an electric-car loan scheme.

The Bikeabout experiment was not specifically designed to diffuse the technology, and both the choice of partners and the EC's funding arrangements reinforced this 'one-off' aspect. As it happened, the project stimulated the interest of one of the suppliers, Dixon-Bate Ltd., which saw Bikeabout as a way to expand its own product market, and the University and the City Council were willing to hand Bikeabout over to them for commercial development.

At present, there are no plans to expand the scope of the Bikeabout scheme to provide general 'public access' to bicycles in Portsmouth. The main difficulty is the need to secure a deposit from the user. If the scheme in its present format were expanded for public use, it is unlikely that the public would be willing to pay the more than £100 it would cost to replace a lost or unreturned bike. Under the present system, with use restricted to staff and students whom the University can track easily, a large deposit is not needed. Money for maintenance is an additional problem. Financial factors were important in the decision not to develop Bikeabout as a public scheme. Perhaps the involvement of a bicycle shop to deal with maintenance could have addressed this difficulty when the scheme was first being designed.

Camden's Accessible Sustainable Transport Integration (ASTI): customizing public transport

Introduction[14]

The UK has an established tradition of local authorities funding community transport (CT) operations to serve people with reduced mobility such as the elderly, disabled people, and children. These operations are part of a range of accessible transport services that include dial-a-ride services, non-urgent transport for hospital patients, and other voluntary and private-sector services. However, inadequate co-ordination of these services currently leads to poor utilization and high cost.[15]

In the early 1990s, Camden Community Transport (CCT) developed its 'PlusBus' project to improve the quality of service and to reduce operating costs by better integrating its accessible transport services. CCT is essentially an established CT operation that uses small diesel minibuses to serve the London Borough of Camden and adjacent areas.

While the PlusBus project was under development, a transport consultant with strong interest in CT and environmental issues suggested to CCT that a high-profile project was needed to increase the public visibility of the relationship between environment and transport. This suggestion eventually led to the creation of the Accessible Sustainable Transport Integration project, or ASTI. 'Accessible' and 'Integration' refer to the strong connection between the project and the development of CCT's community transport, in particular the 'PlusBus' service. The project was thus closely integrated with the development of CCT's operators' network. The trigger for the ASTI project was the opportunity to bid for funding from the European Commission's LIFE programme. CCT and Camden Borough hurriedly put together an initial bid that narrowly failed to be funded, but their second attempt was successful.

ASTI, which ran from late 1994 to late 1997, involved the development and introduction into service of a small fleet of electric minibuses and minibuses fuelled by compressed natural gas that were accessible to all passengers with reduced mobility. The plan also included scheduling-assistance software and vehicle-tracking technologies to optimize the use of these and other minibuses. The cleaner fuels aspect was more important for the Borough than for CCT because of Camden's location in the centre of London where air quality was inevitably a growing concern. But although the project involved the use of cleaner fuel technologies, it was really about 'greening via resource management rather than cleaner fuels', according to one consultant.[16] The major environmental gains came from looking at the resources needed for accessible transport services rather than using the cleaner vehicles.

ASTI benefited from earlier demonstration projects with CNG-fuelled and electric buses as well as from developments in the field of transport telematics.

Other CNG bus trials in the UK have been so successful that the technology is considered to have moved into the commercialization phase. In contrast, only a handful of electric bus projects took place in the 1990s. The most established involved a fleet of four eighteen-seater electric buses placed in service in Oxford in November 1993. These buses connected city-centre destinations with the railway station, where a recharging facility was installed. At first the buses performed reliably and were popular with passengers, but over time battery problems resulted in service reductions, and the fleet was taken out of service in early 1998.[17]

Of the three technologies involved in the ASTI project, the telematics technologies for integrating vehicle positioning, route guidance, and trip scheduling found the widest acceptance. They became diffused within the CT sector and also outside it, and these systems began developing quite aside from ASTI. For example, some providers of accessible transport services began exploring geographical positioning systems (GPS) and the other technologies that were used in the PlusBus service. Mainstream bus operators also began using GPS to combat vandalism and obtain real-time passenger information. Also, although scheduling software for demand-responsive services began to be developed more than fifteen years ago, it is a technology that has matured within the accessible transport service niche. Today such software is tried and tested, several competitors are offering continually upgraded packages, and prices have tumbled.

Objectives and project organization

ASTI was run by Camden Community Transport, with a project manager appointed by the Camden Borough. Overall, ASTI was part of a vision that aimed to provide a better public transport system for people whose options for travel were very limited. The two organizations had a strong pre-existing relationship focused on funding and providing CT services in Camden. ASTI's two prime sponsors, the Borough's Environment Department and CCT, worked together to carry out the vision and to steer the project. They developed the concept, put together bids, and arranged for the involvement of other partners. They also obtained funding from the European Commission's LIFE programme, the UK Department of Trade and Industry, and several private- and public-sector partners (see below). The total funding amounted to nearly 2.5 million Euros.

In detail, ASTI's objectives included:[18]

- Developing a small fleet of three electric and three CNG-fuelled minibuses and the refuelling infrastructure required for these vehicles;
- Integrating satellite vehicle tracking and software for a real-time service optimization and dispatching system to assign trips for a service;

- Identifying and developing the most appropriate routes and services for the vehicles and optimizing vehicle/route combinations for efficient service with minimal environmental impact;
- Improving the quality of life for the elderly, disabled, and infirm by facilitating greater independence;
- Integrating procedural aspects of health, municipal, social, and transport organizations to improve community and health care;
- Disseminating the results of the project so that relevant lessons could be applied across European cities to enhance the provision of transport while mitigating its environmental impacts.

An implicit objective was to develop an environmentally beneficial service using cleaner drive systems and better fleet management, but actual environmental impacts were not monitored.

The vehicle design chosen for ASTI was strongly influenced by the specific function it needed to perform. Vehicles had to offer extremely easy physical access both to ambulatory and wheelchair users. Because no existing electric or CNG buses met this condition, it was decided to use a converted model of the same Iveco Ford van that CCT had been using in a diesel version. This van allowed for combinations of up to 12 seats, could accept wheelchairs, and had a hydraulic passenger lift at the rear entrance. In terms of project management, choosing this vehicle reduced uncertainty and meant that attention could be focused on the scheduling and operational changes required for electric and gas traction.

AC motors were chosen because of their relatively lightweight and sturdy construction. The AC controller developed by the British firm Wavedriver was used; it was actually an integrated AC drive and a high-rate charger and battery-management system. Proven, maintenance-free, lead-acid gel traction batteries were also chosen. With an overall weight of 4.5 tons (including one ton of batteries), the vehicle's range on one charge fell well short of the 130 km required for daily operation. However, a simulation of bus operations suggested that 'opportunity recharging' during occasional layovers between trips could achieve the target of 130 km/day thanks to the Wavedriver system's quick-charge characteristics. For the CNG buses, Iveco Ford, working with the Motor Industry Research Association (MIRA) and British Gas, converted the diesel vehicles to run on CNG.

Evaluation of project design and management

Because ASTI was an experiment involving three inter-related technologies, it created overlapping networks of business, technical, and research partners, along with a network of accessible transport operators to develop the operational side of the PlusBus service. A three-person team consisting of one member each from CCT and the Camden Borough, along with an external consultant, selected the partners; all three were well grounded in the CT world.[19]

The team faxed 100 potential suppliers and received 40 replies. They then carried out a type of competitive selection, but, unlike conventional tendering, this process was intended more to test potential partners' commitment than to push prices down (although prices had to be accommodated within the overall budget). The team then invited preferred suppliers or partners to meet them, visit the site, and make their case regarding what they could bring to the project. A strong commitment to the project and its aims was considered crucial.

Once selected, the partnership network worked together very closely and arrived at joint agreements on key decisions. One aspect of ASTI that was particularly emphasized was the 'openness' of the relationship among the partners in developing the design for the two types of vehicles and the ASTI system as a whole. The consultant commented that the private-sector partners 'got a bit of a shock as they thought the community transport/voluntary sector would be a pushover,' because in practice, they found themselves subject to close scrutiny and binding conditions.

The partner network turned out to be strong, in part because ASTI was important to the partners for developing their core activities. For instance, because MIRA wanted to position itself on the European stage with respect to cleaner fuels, it viewed ASTI as a prestigious project that could establish and enhance its international reputation. For Wavedriver, ASTI provided its first serious on-road project and a chance to gain operating experience with its controller. Given that systems integration in EV design was still underdeveloped, the company hoped to find a large market for its systems. Two other partners were PowerGen, UK's major electricity generator, and London Electricity, an energy supplier, both of which had an interest in developing new markets for electricity and electric appliances for purposes of load management.[20] British Gas, another partner, was also interested in developing new markets.

The team's hardest task was attracting an appropriate software company as a partner in the project. They approached a large number of GPS software companies, but none was particularly keen; they believed that the market for accessible transport was not worth bothering with and that other applications looked more lucrative. Signal Computing, the company that eventually joined the project, initially dismissed ASTI but later decided that accessible transport did represent a market worth developing because success there could lead to applications in mainstream transport.[21]

Among transport operators, the only notable absentee was London

Transport, which is responsible for accessible transport in London. LT did become involved via the follow-on project, PlusBus Interactive. Other than Sainsbury's, the supermarket chain – with whom CCT had already contracted to provide transport services to and from their Camden branch – no other CCT clients chose to join ASTI. The Camden and Islington Health Authority was initially represented by an employee, but this tentative link ceased after the individual involved left. The ASTI partners worked closely with local community and day-care centres through regular user forums and other meetings to ensure that the transport provided was responsive to user needs.

After the project concluded at the end of 1997, its elements became a permanent part of CCT's operations. The next stage is PlusBus Interactive, which studies the possibility of a control centre that allows other transport operators to join in and pool their resources to improve integration and utilization of the shared fleet. This centre would use the scheduling and control systems developed under ASTI. Further European funding was secured for PlusBus Interactive, and London Transport, who did not take part in ASTI, is involved in this service.

Learning

ASTI resulted in a number of useful lessons concerning both vehicle technologies and the operational aspects of an integrated service. Regarding the technology, the experiences were generally positive. The range of the electric buses proved less than anticipated, and this was somewhat aggravated by delays in installing a second charging station.[22] The lower range also appeared to be related to driving styles, and this was addressed by retraining the drivers. The use of lead-acid gel batteries, the Wavedriver system, and the strategy of opportunity charging proved to be a successful combination that met the daily range requirements. Although this strategy had led to cell failures in the Oxford project because the batteries had not been sufficiently discharged, the ASTI buses did not appear to have been affected by this problem. However, heating was a problem on the electric ASTI buses; the batteries did not generate enough surplus heat, and because the users were elderly and disabled, this was a serious issue. As for the CNG buses, fuel economy turned out to be to 20% better than predicted.

With reference to bus technology, ASTI showed that the technologies used were applicable to the CT sector. Representatives of other transport service operators, although impressed by CCT's electric minibuses, still believed that they were too expensive compared to diesel vehicles. They were, however, more hopeful regarding the CNG buses.

The project also established that accessible transport, rather than being merely of peripheral interest, could serve as a staging ground for developing demand-responsive technologies which were then able to diffuse into mainstream transport applications. This lesson was especially important for the telematics firm.

Both drivers and passengers were generally very positive about the buses. Users found both the CNG and electric minibuses quieter than the diesel vehicles they had replaced. This even led to new opportunities for passengers who could now chat with each other. The low emissions were also appreciated; as one CCT driver said:

> As my partner and child have developed asthma, it is a privilege and an honour to drive a vehicle that is, if only in a small way, cutting down on pollution.[23]

The results of the ASTI project have been widely disseminated, particularly in the CT press. Its service-integration aspects were well understood and have attracted wide support within CT groups and the social services departments of the municipalities that fund them.

However, municipal-transport operating departments and others that run accessible services have tended to perceive ASTI as a bus *technology* development project rather than as a bus *system* development project. This perception was illustrated at the mid-project ASTI Technical Seminar organized by MIRA in December 1996. The event attracted a number of non-CT attendees who were interested in applying the ASTI technology to their own organizations. These included representatives from several local authorities and their social services departments, London Transport, and the above mentioned Sainsbury's supermarket chain. The questions asked by these representatives indicated that they generally viewed ASTI as an 'electric bus' or 'gas bus' project. The project's basic vision, of exploring the ways in which information technologies could transform transport service for the mobility-impaired, was not shared and possibly not even grasped.

Among policy-makers, the issue of integrating the operations of parallel fleets of providers, which was central to the ASTI/PlusBus development, became increasingly important after the project. As increasing amounts of money are needed to provide a poor-quality, uncoordinated mix of accessible transport services, local authorities commissioned various studies on integrating ASTI-type innovations with existing transport systems. This interest was not an effect of the ASTI project but was clearly a related development.

Institutional embedding

ASTI contributed to network development by strengthening links between the CT segment of the accessible transport services niche and other providers of accessible transport services. Diffusion of the ASTI fleet-management and trip-allocation technologies has already taken place by mechanisms that have sometimes been surprising. For example, the UK trend towards privatization of municipal services assisted the diffusion of ASTI/PlusBus technologies beyond the community transport sector.

One example is the London Borough of Lambeth, which let out, via competitive bid, the whole of its Direct Works organization including all its education and social-services transport. In partnership with Ealing Community Transport, Camden Community Transport established a new CT group in Lambeth that provides transport services as subcontractors to a large utilities company. A CCT spokesman said:

> Our work on low-emission vehicles, global positioning, and transport management software which we have developed as part of the ASTI project can be applied to the transport problems of Lambeth.[24]

The movement of CT groups and their technologies towards more mainstream transport activities could create competitive pressures on conventional bus companies that may in turn adopt CT-type systems and innovations.

No benefits emerged from the investments made in the six ASTI minibuses beyond their continuing use in the PlusBus Interactive scheme. The electric and CNG buses were one-off conversions to a special design and thus quite costly, for example, about 150,000 Euros for each electric bus. Moving to volume production of twelve or more vehicles of roughly standard design would cut the cost by about half. Beyond this, no economies of scale will be gained until mass production is involved, when orders would have to be in the thousands. Such numbers are not attainable in today's market for accessible vehicles, which is served by fragmented batches of van conversions, but operational integration could lead to more universal specifications, which could in turn result in mass production of accessible, alternative fuel minibuses.

Concerning expectations, the accessible transport services niche aspects of ASTI – i.e., the vision of using IT to integrate poorly coordinated transport for the mobility-impaired – appear to be widely shared. However, ASTI's telematics system was distinctive in that it was not developed simply to improve existing ways of offering accessible demand-responsive services but rather to effect changes in how services were provided. Obviously, political and organizational boundaries become relevant as actors attempt to defend their own empires. For example, under-utilized school transport vehicles are rarely integrated into CT schemes, although they obviously share the roads with CT service vehicles. Additional extensions of the ASTI CT services model to other transport domains are still in their infancy, but most observers expect the continued greening of the CT system via integrative IT improving overall resource use.

The ASTI experiment also helped solidify expectations around specific technological alternatives. For example, a number of technologies can be used to locate the position of a minibus. For demand-responsive services, this has always been an important issue. In cabs, radios have been used since the 1980s, so that if dispatchers want to know a vehicle's precise location, they use the radio to ask the driver. Such cheap, low-tech solutions may well be appropriate for smaller and less complex dial-a-ride and CT operations. But ASTI's use of a fully integrated package of satellite positioning and passenger booking has helped to clarify thinking about what type of system will best meet the needs of both operators and users.

Regarding expectations about CNG and electric-traction technologies, ASTI has apparently had little if any influence. Cost and performance remain the key barriers to electric traction. Electric buses may prove useful for CT services in some cases, but the prevailing opinion is that other alternative fuels (CNG in particular), low-sulphur diesel, or hybrids will be more appropriate solutions.

Reinforced niche development

Overall, ASTI showed that both CNG and electric-bus technologies were viable for CT operations, although, as mentioned above, initial costs of the latter remain a substantial barrier. The alternative-drive technologies used in the project may also find mainstream application in taxi services; because the operational requirements of taxis overlap with those of CT, ASTI's experience with CNG/electric-traction would be relevant. CNG is already in limited use in straight commercial road-transport applications. However, many taxi companies are actively looking at LPG rather than CNG as a cheap, quick way to offer cleaner-fuelled vehicles and thus to keep their vehicles running in city centres where other vehicles could be banned.

ASTI offered positive evidence of the utility of both the concept and the operating economics of the IT software systems and integrated services, and the value of these systems appears to be gaining wide acceptance. The project probably speeded up adoption of GPS and booking/dispatching systems as well as their related software by clearly demonstrating the benefits that can be achieved.

ASTI also showed that accessible transport services are an excellent pioneering application that does not require a large infrastructure investment. The IT technologies developed in the project could find mainstream application in developing general public bus services. Vehicle tracking and real-time information systems are already being introduced in some areas. As new IT technologies are developed, it may be possible to reintroduce dial-a-ride to mainstream operations. These technologies could certainly transform present bus technologies and operational regime and increase the appeal and viability of bus services compared to the private car.

The Swiss mobility co-operative: car-sharing and collective car ownership

Introduction[25]

The two experiments in reorganizing mobility discussed above focused on attracting users to 'environmentally benign' transport modes such as bicycles or buses. Bicycle pools promise to reduce the use of private cars in inner-city areas, while customized bus systems can improve the availability and flexibility of public means of transport.

The individualization of public transport is, however, not the only approach available to encourage a shift to more sustainable transportation regimes. New patterns of car ownership and use hold out considerable potential as well. One of the most promising and fascinating developments in Europe over the past few years has been the emergence and diffusion of organized forms of car-sharing, which severs the relationship between car ownership and car use. Users gain access to vehicles by subscribing to an organization, be it a co-operative, an association, or a firm, that owns and operates a pool of cars. Use is regulated by explicit rules, and members pay for their trips according to their actual use of the system.

The shared use of cars among a limited number of users is nothing new. Members of households routinely share the use of cars or lend them to relatives and friends. Normally one member of the household owns the car and is usually the major user, with use by others and compensation determined informally. Organized forms of car-sharing, as we call it here, differ from private forms because they regulate access, duties, responsibilities, services, and fees in an explicit and professional manner. Furthermore, users gain access not to a single car but to the entire pool of cars owned by the organization. Compared to conventional car-renting firms, organized car-sharing can set up car locations very close to where users live.

Organized forms of car-sharing have been tried and tested for several decades. They started mostly as neighbourhood initiatives, but as such could rarely handle anything more complex than one or two automobiles and a handful of users. This began to change with the emergence of new communication and information technologies and with increasing neighbourhood-level interest in some areas in experimenting with sustainable transport.

One of the earliest examples of organized car-sharing in Europe was the Swiss car-sharing co-operative 'Mobility', founded in the spring of 1997 as a merger of two existing co-operatives, ShareCom and ATG (AutoTeiletGenossenschaft) Schweiz. These two co-operatives had been created independently in 1987 by a handful of households in two different Swiss cities, ShareCom in Zurich and ATG in Lucerne. During the ten years that followed, both co-operatives experienced exponential growth in members, cars, and locations. By 1997, they counted more than 5,000 members each and had established car locations in every major urban centre in Switzerland. By the end of 2000, three years after the merger, Mobility was serving about 40,000 members with 1,500 cars in 900 locations across the country. Mobility co-operates with national and local providers of public transport and is expected to become a major element in any new integrated mobility systems being planned for the future.

The development of the Swiss Mobility co-operative is a model case of an experiment that was initiated and strongly supported by concerned citizens. Both of the original co-operatives started out with only one car, a few users, and the half-stated objective of developing a new, more environmentally benign form of mobility. The members embarked on a development path in which both the early users and the organizations' managers had to learn and adapt quickly. Over time, the users' role changed, and the system became increasingly professional. However, those beginnings had an important impact on the evolution of the current system and on its potential for becoming the core of a more sustainable transportation regime. The following discussion sketches the history of these user-led experiments.

> We said that we did not want increasingly to damage our environment, but we wanted to limit ourselves to a sensible degree. Somehow, we all had some green attitudes, a little bit of it. Not fanatic, not extreme, I found, but just a bit of green consciousness.[26]

This quote by a founding member of one of the car-sharing initiatives indicates the pioneers' mind-set. Many were young people who were fascinated by the idea of sharing consumer goods in general (ShareCom) and cars in particular (ATG). They decided to found a co-operative to bring car-sharing out of the private sphere and make it available to the wider population.[27]

Members raised the capital needed to get started by each purchasing a share of the co-operative. Also, all members committed themselves to carrying out other duties such as vehicle maintenance and repairs, billing, acquiring and introducing new members, etc. Members fulfilled these administrative and operating tasks on a voluntary basis in their spare time. This required high loyalty to the system and was probably a precondition for the co-operatives' health, survival and functioning. Decision-making took place at members' meetings. In both co-operatives, the sole initial technical equipment consisted of one car that was purchased and owned by the co-operative. In the ShareCom case, members reserved the car by calling each other; in the ATG case, they put their name on a 'reservation board' located where the car was kept.

These early users were guided by specific expectations about the future of this technology. They took on their pioneering role in the hope of both improving the technology and demonstrating to others that new mobility forms that could fulfill daily transportation needs were available. The early members were thus willing to provide protection for car-sharing, i.e., they actively defended the system they had created by holding it to a lower standard of vehicle service, which in turn bought time for organizational adaptation and learning. Rather than evaluating the service on the basis of short-term, individual economic costs and benefits, the pioneers engaged in strategic action: they defined their needs in relation to expected societal developments and to future characteristics of the product.

The initial system of voluntary work performed by motivated members worked as long as the co-operative was small because members knew each other and were willing to invest time in booking the car and getting to it. But from the very beginning, both co-operatives rapidly attracted new members and experienced almost parallel growth rates of between 50% and 75% a year. By the end of 1990, each organization had more than 500 members.[28] Over time, new members were less likely to cite ecological reasons for joining; rather, they expressed more pragmatic concerns (car-sharing as the best alternative for satisfying their current mobility needs), and financial motivations (car-sharing as a cheap alternative to car ownership).[29]

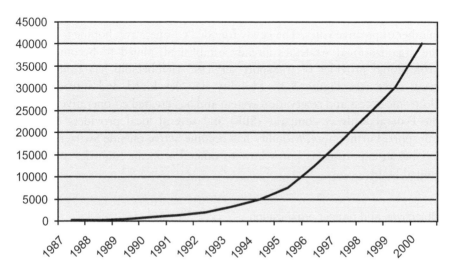

Figure 5.2 Growth in membership rates of Swiss car-sharing organizations; data are for ATG plus ShareCom before 1997, later for Mobility. Data source Harms and Truffer, op. cit.

These changes required new organizational forms and technological systems. Central offices were installed, and a few employees were hired. Local user groups moved to selecting delegates who attended user meetings in their stead. Different organizational philosophies began to emerge. For a long time, ShareCom favoured organic growth and wanted to gain members mainly by word of mouth.[30] ATG, on the other hand, soon abandoned that approach and tried to create growth from the top down by establishing a network of car locations that covered all of Switzerland.

The first step for both in opening the system to a broader membership was to install a metal box containing the reservation book and the car keys at the car locations. But this was only a transitional solution which did not prove well adapted to the demands of growth, so each co-operative, following its respective philosophy, began seeking different technological solutions. ShareCom, hoping to avoid setting up a labour-intensive office, opted for an automated telephone-reservation system. ATG, seeing itself as a service enterprise, opted for a telephone line staffed by an operator. Both organizations kept the car-key box system for several more years and only began replacing it with electronic access cards in 1999. The car-key box system in turn required the installation of on-board computers in the car that automatically registered and transferred all relevant driving data to the central office.[31] Each co-operative invented its own electronic data-processing system for accounting and billing. When Mobility was created in 1997, the merged organization was able to profit from the synergies of the different technological paths that its two predecessors had chosen.

The creation of Mobility was preceded by heated debates within both co-operatives, but ultimately the decision to merge received an overwhelming number of positive votes. The newly founded co-operative, hoping to build on the organizational strategies already established, aimed to become a fully professional provider of transport services. Membership has continued to grow since (see Figure 5.2). Mobility has developed and introduced an electronic access and reservation system and has created co-operative ties with the Federal Railway company (SBB) and several local providers of public transport. Furthermore, Mobility has become active outside Switzerland and has begun exporting its know-how to other European countries and to the United States.

Objectives and project organization

Organized car-sharing in Switzerland was not a formal top-down organized experiment with clearly stated objectives and a formal project structure. Nevertheless, we can identify objectives based on the motivations and expectations of the members and managers who took part in the early initiatives. Their vision was to contribute to the emergence of a more sustainable mode of transportation. To reach this general goal, the groups established a number of specific objectives, including:

- Increasing potential users' awareness of car-sharing by demonstrating its feasibility and cost-effectiveness in everyday use;
- Making car-sharing more attractive to 'normal' drivers by lowering costs and improving the quality of the service;
- Expanding accessibility to wider segments of the population by spreading to new geographical locations and increasing the number of cars available at existing locations;
- Demonstrating to the general population that alternatives to the individual use of private cars exist and that car-sharing and public transport can meet most everyday transport needs.

In the Swiss car-sharing experience, we can distinguish three experimental phases that sequentially led to the fulfillment of the above goals. In the start-up or pioneering phase, the voluntary input of the pioneer members prevailed. The second phase saw the rise of 'competing designs'; both ATG and ShareCom realized the need to professionalize but adopted fundamentally different growth philosophies. The third phase of 'closure' or 'standardization' describes the development of the Mobility co-operative since the 1997 merger and represents the establishment of a professional service firm. For each of these three phases, we may identify an organizational structure, a supporting network, and specific objectives.

From the beginning, the car-sharing experiments were embedded in a wider network of institutions. At first, this network consisted essentially of the households and users who founded the co-operative. Before long, however, new members brought new organizations into contact with the co-operatives, and a number of communities, churches, and political parties supported the expansion of car locations by providing cheap parking lots or by advertising the system among their members.

In the second phase, a more formal network developed. The national traffic club, Verkehrsclub der Schweiz[32] (VCS), gave car-sharing a great deal of publicity in its members' journal and, as early as 1989, developed a manual for promoting private means of car-sharing. Furthermore, VCS helped to build up local user groups in several locations. Finally, and most importantly, it took on the role of spokesperson or go-between for car-sharing organizations within the framework of Energy 2000, a federal research and demonstration programme focused on increasing efficiency in energy consumption. VCS initiated and co-ordinated the first federally-funded promotional projects for car-sharing, and over the years became the main mediator between the car-share managers and government officials.[33]

In addition to supporters, 'competitors' of car-sharing such as the traditional automobile and car-rental industries became involved in Mobility to different degrees. The automobile industry, for instance, observed the developments with interest, and Mobility entered into a wholesale contract with the automobile manufacturer Opel to supply most of the cars in the system. However, Mobility's members did not want closer co-operation with a particular manufacturer because they did not want to constrain future technological choices in any way.

The car-rental firms did not see the car-sharing co-operatives as competitors. Mobility collaborates with some of these firms, but no formal, national collaboration has been established.

Co-operation has been formalized in specific cities such as Zürich, where Mobility formed a joint venture with the local public transport agency (VBZ), the car-rental firm Europcar and the car manufacturer Smart. Within these schemes, each partner offers his own product. This improves access for users, for example in the case of *züri mobil*, where reservations can be made via one common telephone line. In other regions, car renting and car-sharing are offered under the same label and are accessible with one single smartcard. It can be anticipated that in the long run, car-rental firms and Mobility will grow to consider each other as competitors.[34]

Institutional networks also extended beyond national boundaries. In 1991, ATG and StattAuto Berlin founded ECS, European Car-sharing, an umbrella organization for all European car-sharing initiatives.[35] The organization has two main aims:

♦   To facilitate the creation of car-sharing organizations in Europe by disseminating knowledge and information; and

♦   To facilitate cross-memberships and communication about car-sharing by standardizing different car-sharing systems.

By 1998, 40 member organizations had joined ECS representing a total of 36,000 car sharers in eight countries. ShareCom had joined ECS in the beginning but left after several years because it felt that ATG had gained too much influence within the organization. ShareCom never had as strong a European focus as ATG and had chosen to focus on the Swiss market and to rely on an autonomous growth process. Since 1997, Mobility has been a member of ECS.

Project design and management

As stated earlier, organized car-sharing in Switzerland is not a model case for project management because it lacked both a clearly stated set of goals and a formal project organization. Rather, the co-operatives began as small-scale initiatives that over time evolved to manage exponential growth. In just over a decade, the initial experiment changed radically in magnitude and quality. In our interviews, founding members stated time and again that they never would have imagined at the outset where the experiments would ultimately lead.[36]

Although both co-operatives began with no formal project management, managers of both systems were determined to deal with growth in a highly professional way. In the second phase of the experiment, the co-operatives diverged with regard to how formalized and market-oriented the organizations should be, but these differences were resolved during the merger that provided Mobility.

The crucial elements of the learning process are visible when we look at the transition from the pioneering phase to the professionalization phase and observe how well the co-operatives managed the transition. This transition decided the fate of a technological niche that had initially been established by the co-operative actions of a few concerned citizens.

Second-order learning between management and users

One crucial precondition for effective learning was the existence of frequent contact and open communication among members and between members and management. Because of the considerable uncertainty users confronted ('Will I have a car when I need it? What costs will I face afterwards?' and so forth), members needed to communicate with one another about their experiences. Also, the management needed to learn about defects and shortcomings in the system, about the needs of the users, and about required improvements from contacts between and with members. For members, contact with management was important because it reassured them that they were exerting an active influence on the course of the innovation. Specifically, the users played an important role in the development of the car-sharing services by:

♦ Providing 'cheap money' for expanding the system (the co-operative received money from members at low- or even zero-interest rate);

♦ Carrying out tasks such as billing, cleaning, repairing, spreading information, etc., on a voluntary basis, which held down administrative costs.

♦ Publicizing the advantages of the system by word of mouth and thus keeping the costs of expansion low. Members were instrumental in obtaining preferential treatment, such as cheaper parking lots, for the system in their local communities.

Furthermore, they allayed the uncertainties of new members since by recruiting neighbours and friends they gave car-sharing a 'familiar face' and sold the image of a small-scale system despite the co-operatives' exponential growth.

As one former ShareCom manager put it:

> Without the participation, the work of individuals, of members who locally built up the groups, it wouldn't have been possible at all to lead car-sharing, neither ShareCom nor ATG, to the present level. This is indeed the basis, the humus on which all this could be founded.[37]

Over time, as new motivations have became dominant among new members, the role of users as active participants in the system has decreased. Users can now choose between being a co-operative member or a 'mere user'; the second option is a direct response to new users' decreasing interest in becoming actively involved in maintaining the system.

Besides recruiting user support, the management of both co-operatives made conscious efforts to professionalize the organization. They found support early on in Energy 2000, the Swiss government's promotion programme. However, different actors and observers have assessed the importance of such external supports and specific policy programmes quite differently. Commentators from within the car-sharing community have generally maintained that car-sharing would have grown independent of any governmental support, though growth might have been slower and transition periods more difficult. In the eyes of these observers, the essential resources for the development of the system had come from within the organizations and from their members. Many commentators not directly involved in the organizations, on the other hand, have taken the opposite view: that the car-sharing organizations could not have pursued professionalization without 'external' support. Nevertheless, both sides have admitted that the relationship between promoter and promoted has been a process of ongoing learning and trust-building.

The crucial task for all parties involved was the professionalization of the organizations. ATG and ShareCom each tackled the problem quite differently. Despite ShareCom's egalitarian and democratic organizational and product philosophy, management did not hesitate to replace direct communication with computerized systems such as the automated reservation service. ATG, which was more strongly oriented toward a 'business' philosophy, retained direct contact with customers and focused on organizational development. Though the ShareCom members expressed great satisfaction with their organization and with the automated reservation system, the organization also had to face a number of technical problems, especially with its software. ATG's strategy, on the other hand, seemed to show more immediate and tangible results as users maintained personal contact with the central administration.

Policy-makers were eager to bring a more professional attitude and approach into the co-operatives. They claimed to want to get rid of 'ideologies' and pressured both managements to focus on factors such as 'cost structures' and 'pay-back periods'. At the outset, this stance resulted in communication problems between the organizations and the policy-makers involved. In addition, some of the organizations' actual needs became clear only over time. After the first contacts had been established and both sides had gained concrete experience, however, the mutual trust and understanding that grew up between the co-operatives and the policy-makers greatly enhanced the overall responsiveness of both realms to the changing realities of the car-sharing experiments.

It was also concluded that policy programmes could not do everything. For instance, the managements stated that it was not the policy-makers' responsibility to prevent management from learning processes that might sometimes be painful. Rather, policy-makers were to play a supportive role at best. The policy-makers in turn stressed management's responsibility as a key variable that would determine success and limited their own role to addressing major bottlenecks that were beyond the purview of the individual managers.

VCS played a crucial role by building up legitimacy for the car-sharing organizations and establishing a 'translation interface' between the two co-operatives on the one hand and between the co-operatives and the policy-makers on the other. Very probably the two co-operatives would have found it difficult to deal with the support infrastructure without VCS as an intermediary partner.[38]

Surprisingly, both the promoters and the promoted agreed unanimously that direct funding was unnecessary and could perhaps be harmful. One important argument presented by the co-operatives' management was that car-sharing still represented a predominantly private form of transport based on motorcar technology. Because the co-operatives did not want to promote additional automobile use, they argued that users should carry the full costs of their consumptive act of car driving individually. Direct subsidies to the car-sharing organizations might have made the use of cars too cheap and thus encouraged users to drive more than absolutely necessary.

The majority of the interviewees inside and outside the car-sharing co-operatives felt that financial support was not lacking for specific projects. Even so, managers stated that large injections of funding might have posed a considerable threat to rapidly growing organizations like theirs. As one manager put it:

We coined the term 'subsidizing to death' with regard to governmental support projects. The problem with these clearly defined projects, which run over a period of two to three years, is that they are financed only in part by government programmes.

> We had to bring up the rest. We were thus induced to undertake much riskier projects than we would have done otherwise . . . Sometimes we feared that our decision to start such a project might break us.

Another manager stated:

> The bigger the partner and the project, the more expensive it is for our organization and the smaller the return. The big projects increase the pressure for professionalization, but they tie up resources that are lacking later on when they should be available to resolve more pressing problems.

Many of the respondents said that because problems changed extremely quickly, given the sector's annual growth rates of more than 50%, the key to success was to have a flexible scheme of promotional activities.

Second-order learning among users

The Swiss experiment was a continuous learning process that included first- and second-order learning at the level of management and its interaction with users and policy-makers, where assumptions about managing the rapid growth of car-sharing co-operatives were questioned and revised.

But second-order learning processes were not limited to management. Users also learned much about their daily transport needs and how they could satisfy them by using the new technology. The following is a typical comment by a member who owned a car before joining one of the co-operatives:

> *Member*: [The privately owned car] stands in front of the door. You can get into it at any time, and I think I drove quite a lot [at that time], for example, going quickly here and quickly there – which I don't do at all now.

> *Interviewer*: And which you don't miss either?

> *Member*: No.[39]

When drivers' own cars are immediately available at all times, drivers will tend to use them out of habit, without much conscious decision-making.[40] Car-sharing cars, on the other hand, must be reserved in advance, so users have to plan their trips if the car is to be returned within the time span allotted to their reservations. This system makes it all but impossible for drivers to take the car for granted.

The need to plan trips in advance places car-sharing and public transport on an approximately equal plane. In both cases, users have to accept the short walk required to reach the collection point. The costs of car-sharing and public transport become almost equal as well. Because the car-sharing fees include all fixed costs, using a shared car appears expensive compared to public transport, which users had typically viewed as much dearer than driving prior to joining the car-sharing system. Public transport thus became appealing, and users tended to test it out and then use it more frequently. Users also learned about the advantages and disadvantages of each means of transport and began to use the car only in cases where public transport was not suitable or available. As one former ATG-member put it:

> In time, I had to say, it is not bad, riding a train. Due to specific experiences, I came to the conclusion that you learn to value travelling by train. You come to establish a 'natural' relationship with the car.[41]

Car sharers were much less 'auto-mobile' than car owners. Car sharers who were former car owners already drove less than average Swiss car owners before joining the system (about 9,300 km per year compared to a mean of 13,000 km).[42]   After becoming car-sharing members, annual kilometres travelled dropped to an average of 2,600 km. Some used public transport, bicycle, or motorbike (about 4,000 km) to make trips that would formerly have been made by car, but a considerable number of trips were simply not made. Also, formerly car-free households, who used occasionally to borrow or rent a car, did not drive more after joining a car-sharing organization.

Changes in mobility patterns are initiated not only by the car-sharing system but also by changes in life situations, such as when users inevitably move to other towns with different private and public transport conditions, get new jobs in different places or with different working conditions, or experience changes in income. Change can also come about when a user's car breaks down or when a series of latent experiences accumulates and, at a certain point, reaches a critical threshold that forces a change; examples might include increasing difficulties finding parking, longer commutes due to traffic jams, or increasing car-repair costs. Latent experiences that result in changes in mobility patterns can also be more indirect, for example, if drivers become aware of the ecological problems associated with car use and of the contradictions between their knowledge and attitudes on the one hand, and their behaviour on the other. Such changes can bring consumers to a conscious decision point and result in a break with existing mobility habits. Consumers may then re-evaluate their current mobility pattern and open their minds to other options.[43]

Despite the car-sharers' reduction in total mobility, they did not feel restricted. In fact, they reported an improved quality of life. This might be because in many cases their general lifestyle changed in the wake of their joining the co-operative, but it might also be an attitude change brought about by positive experiences with car-sharing. Many car-sharers reported that they experienced car driving as more and more stressful, and public transport as a luxury since not being behind the wheel meant that they were free to read, sleep, or chat with fellow passengers. In some cases, giving up 'unnecessary' car trips was accompanied by a change in other activities as users reported seeking out shopping and leisure-time opportunities closer to home.[44]

The most easily identified changes in mobility behaviour were observed among car-sharers who had owned a car before joining the co-operatives. Formerly car-free households were already accustomed to coping with the lack of instant availability of cars and demonstrated public transport-oriented mobility patterns prior to joining the car-sharing co-operative.

For these households, car-sharing signifies an increase in quality of life. When borrowing cars from others is no longer possible, quality of life suffers, and the perceived need to purchase a car of one's own grows. Car-sharing is a good complement to public transport-dominated mobility patterns, but since it is only attractive under certain life conditions, where specific learning and experience has led the potential new co-operative member actually to join the service. In other cases, if car-sharing no longer fulfills individual mobility needs (for example, if a family moves to a small village where car-sharing is not available) and a private car must be purchased, the new car owners will fall back into their former use pattern, and the car will come to dominate the household's transportation routines once again. Learning processes are thus reversible.

In sum, car-sharing can be considered a useful means for initiating a transition to public transport-oriented forms of mobility or as a means of stabilizing such mobility forms. In many instances, former car owners have been shown to undergo substantial second-order learning processes regarding mobility behaviour and attitudes. And car-sharing certainly assists formerly car-free households in maintaining their reduced mileage, and individual, motorized mobility patterns.

Institutional embedding

The success of organized car-sharing is strongly dependent on adequate network management. As was described above, the network of people and institutions involved in the car-sharing co-operatives grew gradually over time. A major boost in professionalization resulted from the creation of a translation interface between the two competing co-operatives and between them and the policy-makers. During this period, the focus was on improving car-sharing as such.

Later on, integrating car-sharing into a broader concept of mobility alternatives became more important. The co-operation of Mobility with Zurich's public transport agency was one of the most notable activities in this area; Mobility has been part of this combined mobility system (*züri mobil*) since February 1997.[45] Another example is the joint venture with the Swiss

Federal Railway company (SBB), which offers a combined train-car-sharing ticket. SBB provided car-sharing locations for Mobility at its major train stations, which, in turn, multiplied the number of car-sharing locations.

Furthermore, government-supported network formation and societal embedding resulted from the funding of research projects on the environmental benefits and market potential of car-sharing. These projects helped to diffuse information in specific actor fields and legitimized the various organizations. Car-sharing also became an appealing topic for the local and national press and electronic media.[46]

Of key importance for the development of the networks was the input of specific individuals, their willingness to take risks, and their flexibility in embarking on learning processes. Their versatility was reflected in the changing roles some played as the car-sharing co-operatives developed. The key person at VCS, for instance, left it to become head of a newly founded car-sharing organization, and later became a member of Mobility's governing board. An official of Energy 2000 who was in contact with the car-sharers later started working on the *züri mobil* project. Another transport researcher who was among the founders of ATG later worked for Mobility as a consultant and finally went to the US to work as a car-sharing consultant.

Most participants in the co-operatives mentioned networking and the creation of legitimacy as crucial. These factors helped to fight the negative image attached to most bottom-up initiatives, which tend to be seen as ideologically motivated and un-professional regardless of their actual motivations and capabilities.

Finally, societal embedding of the car-sharing option had to take the international dimension into account. From 1997 on, Mobility was very active in establishing links to other European car-sharing organizations. Management adopted an explicit strategy of contributing to the establishment of a car-sharing standard in Europe. This opens the possibility of considerable market growth. In Germany, in particular, integrating local car-sharing initiatives could in the long run lead to a market boom of considerable magnitude.

Societal embedding also took place with regard to the alignment of expectations about this new technological option. Car-sharing was originally considered as an option for households that could not afford to own a car or for people with strong ideological convictions about environmental issues. But car-sharing is also a mobility option with a distinct market share and serves as a crucial complement to public transport. The Swiss car-sharing co-operatives were able to demonstrate that ownership and use of automobiles can indeed be separated without much reduction in comfort or convenience.

At the technological and infrastructural level, car-sharing initiatives became a testing platform for new software to handle vehicle access and tracking.

These could form the basis for further alignment and integration of different means of transport, such as universal systems for access to public and semi-public means of transport, automatic billing and information services, and the like. From this point of view, car-sharing may well become an enabling technology that encourages future integration of mobility services, and therefore it can be seen as an essential step on the path toward a more sustainable transport system.

A promising niche

Car-sharing offers great potential for achieving more sustainable mobility practices. But is it also a model case for Strategic Niche Management? The Swiss initiatives were not formal research projects or studies. No explicit experimental design, based on a formal analysis of the technological development potentials of different transport innovations, was articulated. Rather, the co-operatives were created by ordinary citizens who shared a vision of a better life and the conviction to reach that state. In the beginning, the initiatives were 'neighbourhood projects'. Later they evolved into full-blown businesses. Nevertheless, we would argue that the co-operatives are in fact a model case for SNM.

The essential criterion is not the existence of a formal experiment, but rather whether the activities focused on creating and developing a technological niche. Niches are protected spaces for immature technologies that are built up by specific actors with a strategic interest in promoting the technology. All these elements were present in the car-sharing case.

Protection for the immature technology was provided first by the early user-managers and later by new user groups, which adapted their expectations to compensate for the system's early shortcomings. Over time and with market expansion, the need for this kind of protection gradually decreased, and the organizations had to cope with normal user expectations and full-cost pricing. The example thus illustrates the different development stages of a new technology, starting from early design variations to the selection among competing designs, and to the upgrading of a technological niche into a market niche.

Network management and societal embedding had been among the most crucial resources required for the initiative to flourish. Supporting networks changed as the markets expanded. Meanwhile, some former supporters turned into opponents, and more professional actors entered the stage, as the technology matured. Furthermore, the 'experiment' of organized car-sharing stopped short of promoting a specific design, i.e. the particular design of car-sharing that had currently developed in Switzerland, but has the potential to encourage niche development across a broad range of potential fields of application.

First and second order learning have been essential components of the experiment as well. These multiple levels of learning were enabled by the close interaction between developers and users, by the consistent support of governmental policies, and by the interactions of different actors from competing and/or complementary product fields such as the car-rental industry, public transport, etc. Learning was not restricted to management but also encompassed customers' growing awareness of their own mobility patterns.

One remarkable characteristic of car-sharing at the consumer level was that it helped to break up user routines by transparently disclosing information about different mobility alternatives and undermining the entrenched linkage between car ownership and use and associated feelings of comfort and security. These characteristics led to consumer empowerment and to the unlearning of environmentally destructive routines and convictions. It is in this respect that car-sharing may deliver on its truly revolutionary potential.

All these positive assessments do not, however, lead to the conclusion that car-sharing as such will lead to a sustainable transport system. Market shares in many countries are still negligible, and the potential for future expansion is difficult to predict. Many more innovations will be needed to achieve a truly long-term system of sustainable transportation that will deliver access and mobility to a majority of the world's citizens at a low environmental cost. Car-sharing may prove to be only the first step in a direction that holds out promise for future development. However, in our analysis it definitely looks like that first step. Let us now turn to the different starting point sketched in the final case study.

Praxitèle: individualized public transport in France

Introduction[47]

Like car-sharing, an individualized public transport system could alleviate traffic and parking congestion and the accompanying air and noise pollution. Such a system consists of individual vehicles that can be rented on a short-term basis (as short as one hour) from a series of stations. Such a system differs from car-sharing in two major respects. First, public, self-service, rent-a-car systems provide vehicles that can be used without advanced reservation, whereas car-sharing is available only to members of the co-operative who must reserve cars in advance. Second, car-sharing members must return vehicles to the location from which they took them, whereas the idea of self-service cars is that they could be left anywhere that was convenient to the user.

The concept is not new. In the 1970s and 1980s, a number of experiments in individualized public transport were held in several European and US cities such as Montpellier, Amsterdam, Brussels, West Lafayette (Indiana), and San Francisco.[48]

These projects, all of which failed for various reasons, have nonetheless produced valuable lessons for succeeding projects. It was learned, for example, that any individual public transport system must have enough vehicles so that users need not wait too long until one becomes available. Also, the ability to track vehicle location is important for managing demand and to reduce the risks of theft and abandonment. The system must have the capacity to allocate and move available cars from stations where demand is low to high-demand stations so that potential users will not be disappointed because no vehicle is available. Self-service systems also require easy, accurate automated payment systems. Furthermore, the interests of taxi companies in the areas served by self-service systems must be taken into account to minimize what might be perceived as unfair (that is, subsidized) competition. Self-service systems also encountered some of the typical problems experienced by car-rental companies, such as vandalism and excessive wear-and-tear on the vehicles. Rental companies usually keep vehicles less than six months because so many clients drive the cars, mileage accumulates quickly, and maintenance is often neglected. Finally, earlier experiments that used EVs suffered from those vehicles' poor performance and contributed to the rejection of those systems.[49]

Interest in the idea of individualized public transport was reborn in France in the early 1990s, when the environmental effects of transport and traffic congestion began receiving much attention and a number of new technologies were developed that promised to solve the previous problems with self-service systems. These technologies included inexpensive methods of communication and locating and tracking vehicles, computer programmes for fleet management, smartcards for payment and easy access to vehicles, and a new generation of EVs.

In 1990, researchers connected with the French bus and railways operator CGFTE organized a series of meetings to involve passengers in the planning of future transport services.[50] Passengers expressed a strong preference for a self-service system based on rental cars because it was believed that this would combine the advantages of private cars and public transport. Such a system, the passengers said, should guarantee parking spaces for car users, be easy and convenient to use, and employ non-polluting vehicles.

When the researchers, who were surprised by these responses, presented them to CGFTE management, they found that the company was already interested in developing the idea. CGFTE had been seeking ways to make its public transport service more flexible and better aligned with individual needs. The company believed that transport systems of the future should include a range of co-ordinated transport modalities, among them self-service vehicle-rental schemes. CGFTE decided to use EVs because the technology was becoming popular in France at the time and using it would heighten the company's profile with customers as well as with the government and Electricité de France (EDF), the French public electric utility. In 1991, EDF and the Ministry of Transport decided to support a small-scale feasibility study of an urban-transport system based on self-service EVs.

Concurrently, two national research institutes, INRIA (information technology and automation) and INRETS (traffic systems),[51] launched a programme to promote an innovative transport system based on self-service EVs.[52] In this case, the focus was on off-road applications such as shopping malls, airports, festival grounds, etc., and one important aim was to develop both semi-automated and fully automated vehicles. The programme aimed to undertake an in-depth analysis of the whole range of technical and organizational problems and to take part in experiments with EVs. The programme also hoped to evaluate the societal conditions needed for such plans to succeed. In 1992 the CGFTE and INRETS/INRIA programmes merged, the Praxitèle consortium was formed, and the suburban Paris town of Saint-Quentin-en-Yvelines declared its interest in taking part in the experiment.

The consortium commissioned several studies. The above-mentioned feasibility study investigated previous projects and formulated recommendations that it was hoped would avoid repeating earlier problems. A marketing study concluded that there was a demand for small self-service vehicles. A survey to assess public reactions to the Praxitèle concept registered a high level of understanding and positive expectations among interviewees: 7–11% of individual car users in towns were willing to use the Praxitèle service for 42–45% of their trips. This implied that the service could replace 3–5% of trips currently being made by car in urban areas.[53] Four potential customer target groups were identified:

1   Professionals, including executives and other working people aged 35 to 65 who were expected to be interested because the system would save time and make parking easier. These drivers might use the rental vehicles on a daily basis instead of using their own private car;

2   Drivers who might give up using their own cars and rent the EVs for private trips to solve parking problems;

3   Executives who would use the EVs as alternatives to taxi rides and travel on public transport to attend meetings or make short professional trips;

4   Tourists who would use the EVs for sightseeing.

A series of other studies carried out between 1992 and 1997 analysed the Praxitèle system's social and economic feasibility, its organization, and the experiment's design and setup in Saint-Quentin-en-Yvelines. The consortium also constructed a miniature 'demonstrator' consisting of two electric Renault Clios equipped with sophisticated information and communications technology (ICT) functions, two stations with inductive charging facilities, and a central station to manage the 'fleet'. For some time, the main bottleneck appeared to be securing funding for the project, but by 1997 the consortium had obtained support from the European Commission, and this, together with funding from a number of French sources, allowed the project to begin.

Objectives and project organization

The Praxitèle experiment's overall objective was to demonstrate the usefulness and economic feasibility of an individualized public transport system based on a centrally managed fleet of small EVs. The evaluation focused on the following aspects:[54]

1   The technical adequacy of the vehicles and their related infrastructure; this included testing the new inductive recharging system;

2   How well the technical system's usefulness and reliability met customer requirements;

3   Public acceptance of the new service;

4   Conditions that would maximize use of the system, including fares, hours of service, and other factors;

5   The system's organization and operation;

6   Knowledge required to expand the system to other sites.

A number of partner companies took part in setting up and carrying out the experiment. Transport operator CGEA (previously CGFTE) led the project and provided financial, commercial, and organizational management and co-ordination.[55] EDF designed the battery-recharging stations and managed power distribution. Renault designed and manufactured the EVs. Dassault Électronique developed the on-board data processing and communication systems. Finally, INRIA and INRETS provided research and technical-scientific support.

The pilot scheme had a budget of 30 million French francs (4.6 million Euro). The companies and Saint-Quentin-en-Yvelines provided half, with the other half coming from the national government, regional public bodies, RATP (the Paris transport authority), and the European Commission.

Praxitèle was based on a fleet of *Praxicars*, small EVs located in specific *Praxiparcs* and supervised by the *Praxicentre*, a central computer. The experiment started in October 1997 in Saint-Quentin-en-Yvelines, which is located about 15 km west of the heart of Paris and has 140,000 inhabitants. It has two expanding industrial zones and is considered the second most important business area in the west Paris region.

Fifty Renault Clios with a 70-km range were available from fifteen (expanded from the original five) stations located strategically within the town at railway and bus stations, shopping and business centres, hospitals, etc. The choice of the vehicles was a compromise in that the Praxitèle system intended to use small and relatively cheap EVs for short-range trips. Because such vehicles were not available on the market, however, the project used the relatively heavy and expensive Clios, which Renault was producing in sufficient numbers. Because a major focus of the project was to evaluate user acceptance, rental prices were subsidized so that users paid only the price that would have been set if cheaper EVs had been available.

Users could take any free EV at any time from any of the stations, use it freely as if it were a private car, and then return it to any of the stations. Several drivers could use the same car during the day, thus theoretically reducing the total amount of parking space required in the city centre. Drivers had to possess a valid driving licence and to register as a member the first time they used the system. From then on, they could use the car as they liked, much as they would have used their own private vehicle. At the end of each month, customers were billed for the time they actually used the service.

Computer terminals installed at the stations managed recharging as well as billing, relevant fare schedules, and information on vehicle availability. The terminals also provided information on other public and private transport options such as taxis and allowed users to order taxis if no Praxicars were available. EDF designed and developed a new inductive recharging system for the EVs' batteries. This meant that there was no physical contact between the vehicle and the charger, thereby minimizing danger to users who might be unfamiliar with EV technology.

Users gained access to the cars by using the *Praxicard*, an electronic credit-card-sized smartcard that functioned as an entry key and also started the motor. The Praxicard connected to an electronic billing system identical to those now in use by other public transport services and could also be used to reserve vehicles via computer terminal or by telephone.

The central computer system controlled the fleet of cars and could inform users of the nearest available vehicle via computer terminal or by telephone. The computer centralized information about vehicle condition (for example, state of charge), the availability of parking spaces, and user accounts. Telecommunication links to the vehicles were used to authorize car access, calculate the fare (preferential rates were available depending on the time of use and the location of the vehicles), redistribute empty cars among the Praxiparcs, and decide whether vehicles needed recharging or not. The network also handled reservations and transfers to other modes of transport such as train, bus, or taxi.

In the first version of the Praxitèle system, users drove the vehicles but future versions hold out the potential of partial or even fully automated driving, with passengers travelling on dedicated tracks of an automated highway. INRIA in particular is working on this aspect.

Project design and management

Preparation of the Praxitèle experiment took five years, a relatively long time, for two reasons: first, the complexity of technologies involved, and, second, difficulties in finding financial support due to the public authorities' reluctance to fund risky projects.

The Praxitèle system involved a range of innovations employing non-standard technology. Neither users nor operators were familiar with many of the novel systems. The combination of the technical innovations and the concept of self-service EV cars, a radical and innovative concept in transport technology, resulted in a project that ran a high risk of failure and raised problems about how to assess user acceptance and the ultimate potential of the system.

The project organizers were well aware of the risks involved and therefore, although they were among the largest companies in France, sought public support. However, public authorities were reluctant to provide funding precisely because of the perceived risks. The project was delayed for almost two years while the proponents tried to secure support. They eventually succeeded, in part, thanks to several studies that increased outside interest in the project.

The initiators had no alternative but to use real-life experimentation because the technology was not available anywhere else and the concept had never been tested in this way. Also, because not all the technologies involved were at the same level of development, two different operating modes were chosen for the experiment: the first required the system operator to intervene; a later implementation allowed complete self-service. Technical reports prepared during the project concluded that future applications would need to be more reliable and easier to use.

Learning

The project managers developed numerous monitoring systems to learn about the functioning of the technology, use patterns, and user acceptance. Preliminary results clearly indicate that users have been very satisfied with the service. Users were especially appreciative of the self-service system's freedom and convenience and their ability to ignore the issue of vehicle maintenance. Judging by the low incidence of accidents and vandalism, users have driven responsibly and treated the vehicles with care. Users also praised the use of EVs for environmental reasons. They generally use the cars only as the study intended, that is, for short trips between the dedicated stations in Saint-Quentin-en-Yvelines.

The number of people using the system has grown during the project. In May 1998, 400 members had registered; a year later, this number had doubled.[56] In the first eighteen months of the experiment, users made 25,000 trips that covered an average distance of 8 kilometres in 15 minutes. By April and May 1999, average use was about 500 trips a week, or about one and a half trips per car per day. Three out of four trips in March 1999 originated from or ended at the rapid suburban railway (RER) station that connects Saint-Quentin-en-Yvelines with Paris. This shows that the service was used mainly to supplement public transport.

Two out of five users declared that if the Praxitèle service were not available, they would have used their private cars, while the remaining three-fifths would have taken the bus. Moreover, the frequency of Praxitèle use increased at hours when fewer local public buses were running. Relatively few people said they would not have made trips had the self-service system not been available.

In the course of the project, more or less stable patterns of journeys among the stations appeared. For example, in April 1999, 95% of the trips were made between seven of the stations and 33 cars would have been enough to satisfy demand. Such data are essential for the eventual commercialization of the system because they allow cost-effective planning. Typical users tended to be males (74%), employed (90%), and mainly in management positions (50%). Over half of the users were aged 35 or younger. Four out of five users lived in the town itself, and 62% lived within 400 metres of a Praxitèle station. The strongest predictors of use seemed to be the availability of a private car and the distance from the user's home to the station. The most frequent users were those who did not usually have a car at their disposal, either because they did not own their own car (26%) or because they shared it with other family members (29%); distance to the station was not very important for these users. The service's main advantage to users without their own cars was that it made them less dependent on regular public transport and on scheduling the use of the one family car.

The 44% of Praxitèle users who owned private cars used the system only if they lived close to a station. To them, the service appeared practical only if their own cars were temporarily unavailable; also, using the service guaranteed them parking at the RER station. The few companies affiliated with the project reported that the service met employee travel needs during the workday, especially for errands to the bank, post office, or city administrative offices.

These findings suggest that a major selling point for such a system is its convenience as a supplemental transport service for both owners and non-owners of cars. In addition, the service can be offered as a supplement to collective public transport and as a replacement for bus services where such service is lacking or at times when buses are infrequent.

When non-users of the service were studied, interviews revealed several principal barriers to adoption. First, many non-users did not perceive a need for Praxitèle because they were satisfied with their existing transport options. Second, they lacked information and did not realize that it was a self-service system; many said they thought that it was basically a technical test. Third, many expressed a lack of curiosity concerning innovations in transport. Fourth, non-users often said they would find it strange not to own the car they used. Companies that failed to use the system reported reluctance to consider it because it was only experimental and might not continue to be available over the long term. They therefore did not try to learn about the service, distribute information to their employees, or make user passes available to employees.

Most other problems and issues related to the service's experimental and innovative features. During the experiment, continuous attempts and adaptations were made to solve technical problems. For example, the effectiveness of the inductive charging system proved to depend on how exactly the cars were aligned with the charging stations. Inexperienced users often needed the assistance of staff to achieve the right alignment between the coil underneath the car and the one in the station floor. Over time, the project managers decided that inductive charging discouraged use, so it was abandoned in favour of cable recharging.

Another discovery was that actual 'self-service' was crucial. User demand for cars increased markedly after the initial operator-assisted system was replaced by round-the-clock unattended renting. Use by private individuals saw a particularly noticeable upturn, while companies generally requested a reservation system to guarantee vehicle availability.

It was evident from the beginning that under current conditions a self-service system would not be economically feasible. EVs are far too expensive because they are not mass-produced, and an economic break-even point could not be achieved for any plan that used fewer than several hundred vehicles. However, the experiment has proved that there is a substantial demand for a Praxitèle-type service. How this demand could be satisfied in a cost-effective way remains unclear in the absence of detailed economic evaluations.

The system's environmental impacts were not analysed. Moreover, because Saint-Quentin-en-Yvelines is relatively sparsely populated given its geographical area, availability of parking is not a pressing issue. Thus, it was not possible to assess parking congestion as an incentive for using the system.

Institutional embedding

Establishing the legitimacy and effectiveness of the Praxitèle service as a public transport service in conjunction with collective transport and mass transit is believed crucial to its future. One objective of the experiment was to integrate Praxitèle into the existing transport system of Île-de-Paris by use of a common, Paris transport-approved pass. However, this was not achieved due to technical problems and delays.

At first, there was concern that Praxitèle would compete directly with taxis and public transport services. Thus, the Paris transport authority RATP was reluctant to support and co-operate in the project. When the project began, the taxi driver co-operatives in Saint-Quentin-en-Yvelines threatened to protest because they feared that they would lose customers. However, this did not happen because Praxitèle mainly attracted users of private cars and buses. In fact, the taxis actually increased their customer base because the Praxitèle computer terminals offered users the option of calling for a taxi if no self-service cars were available.

The results of the Praxitèle experiment have been followed closely by other parties interested in self-service systems. As was discussed extensively in Chapter 4, for example, the competing French Liselec consortium, an outgrowth from Peugeot-Citroën, has developed the TULIP concept and has been testing the operation of self-service electric cars in La Rochelle. The French Ministry of Transport tried to foster co-operation between the Praxitèle and Liselec consortia, but apparently did not succeed.

In Switzerland, meanwhile, the CityCar project in the town of Martigny provides 30 lightweight self-service EVs for public use. The objective of this project was to pilot-test the transport concept rather than carrying out a market test. Palermo, Italy, and Tokyo and Yokohama in Japan are among several other cities where car-sharing projects using EVs have been started.

The Praxitèle experiment involved a network of large companies whose combined efforts could have mobilized considerable resources; but in practice, the project remained rather marginal. The partners paid little attention to public relations at the start of the project because they just wanted to conduct the experiment rather than turning it into a special event. Thus, no funds were set aside to promote the launch of the Praxitèle system and, when the consortium realized its omission, it merely hired two students to organize a publicity campaign.

One reason for this lack of attention to publicity was that the companies involved were accustomed to working in highly competitive markets where secrecy is a competitive weapon. In addition, the electronics companies often worked in the defence sector where covert projects are the rule. The existence of the competing Liselec project (see above, and also Chapter 4) and the delayed start of the experiment also prompted the partners to maintain a low profile. More importantly, the experiment failed to become widely embedded within the partner organizations. The motivation and involvement of the

CGFTE project team was not mirrored in the wider CGEA organization. This caused confusion as to whether the project was a test of an innovative system, and thus focused on learning about the system, or whether it was a test of a transport system that needed to show profitability quickly. Nevertheless, the CGEA management had a keen interest in the system and now plans to repeat the experiment elsewhere. The participation of the wider Renault organization was also minimal. The company seems to have been interested in Praxitèle mainly as an avenue for selling its EVs.

The project confirmed expectations that an individualized EV-based public transport system satisfied a need and could work. Whether it is also economically feasible remains to be seen; testing economic viability was not part of the experiment. Contrary to expectations, the majority of users were not employees of local companies or tourists but rather residents in the area who added the service to their spectrum of transport options as a convenience. More employees might have used the system if companies had distributed information and passes; this would ideally become the focus for a follow-up experiment.

Creating a niche

The Saint-Quentin-en-Yvelines experiment seems to demonstrate that there is a place for a self-service car system in urban transport. The project managers are satisfied that the technical feasibility of the self-service system has been proven and that the final results will help to define the conditions for successful commercialization of the service.

Three crucial factors for future commercialization will be: enough vehicles and stations to meet the needs of the relevant population, good user information on vehicle availability and the status of the EVs' batteries, and an appropriate balance between an entirely automated self-service and tangible, visible human staffing. The latter is necessary because of user concerns about possible breakdowns of the complex system, and attendant safety issues.

Larger-scale projects are being considered for other sites, the most far-reaching of which would be a project for Paris involving some 2,000 cars. Economic studies indicate that self-service transport systems can be profitable but require large upfront investments. For such systems to succeed, it is essential to identify operators who are willing to take the financial risk of approaching cities or public transport authorities and who are also willing to set up and run the complete service. CGEA may be such an operator; the chairman of the CGEA subsidiary that works on urban transportation systems was quoted in January 2000 as follows:

> We are working on new applications that are simpler and more commercial. During the current year, we are going to launch a second-generation Praxitèle service at a

number of sites in France. There will be new electric cars with better performance and a longer range. We'll make a special effort to reach business customers [such as] the employees and business visitors we dealt with in Saint-Quentin-en-Yvelines.[57]

Experimental findings

Contributions to niche development

When it comes to niches, the Bikeabout case showed that, apart from acting as a breeding ground, niches can also become prisons for a technology. Due to specific contingencies and conditions, contacts between the Portsmouth City Council and a University trainee resulted in an effective bicycle-pool scheme. Applying information technologies could now solve a number of problems that similar schemes had faced in the past. The scheme was successful in terms of its own aims, and it worked without many problems. However, except for the supplier of the electronically controlled racks, the actors involved did not have a stake in diffusing the scheme. They learned how to operate and use it, but this practical experience did not advance learning about possibilities for reconfiguring of mobility. This shows that providing an easily accessible and workable technical alternative is not enough to change mobility patterns. To establish bicycle pools as a real mobility solution, this niche must be replicated in a context where the scheme can be combined with strong incentives or even legislation to get people out of their cars.

The ASTI project was complex, involving CNG and electric buses as well as transport telematics for vehicle positioning, route planning, and trip scheduling. Its underlying objective was to integrate disparate public transportation services for the elderly, the disabled, and children. Various views on the project developed among the partners. It was seen as a niche for testing CNG and EV buses but also as an experiment and niche for upgrading public transportation to make it more attractive. The integration of services, the development of dial-a-ride, and the positive feelings evoked among users by the EV buses could have been harnessed to impel additional experimentation and systemic change. The testing of the technologies and the development of integration measures were successfully completed and are irreversible. ASTI was part of a process of niche branching through which the various technologies involved established themselves. Yet here again we see the niche operating as a prison. Other CT providers adopted ASTI's integrative approach only patchily, and mainstream public transport providers have also failed to embrace the vision of integration. How to diffuse the ASTI experience and its technologies to public transport therefore remains unclear.

The car-sharing history is quite another story, for it has become a mature mode of transport in Switzerland and promises to be established in other countries as well. The technological niche expanded into a thriving market niche during the 1990s. Its institutional embedding is strong, and the network behind it will not easily disappear. Complementary technologies have been developed, substantial investments have been and are being made, and expectations for the future are robust and broadly shared among the managers, users, and government agencies involved. The car-sharing case is the only one of the four in which we can see mobility being reconfigured. System users were led to rethink and change their mobility patterns, which resulted in a reduction in car travel. This case shows that users can change their perception of the advantages and costs of owning and using private cars and transform their conventional use and ownership patterns. The users' participation led to their unlearning and abandoning their habitual mobility patterns.

How can we explain this success? We suggest that the intense interaction between the users and the managers of the car-sharing business was crucial. This interaction was in some ways more like a fusion in that the users initially played an important role in running the system. This case shows that user adoptions of innovations follow not only from *de facto* use and objective user characteristics but also from the quality of interactions between producers, users, and any third parties involved. Whether innovation will be supported by such interactions depends on the learning environment created in the innovation process. A good learning environment provides, in addition to learning by doing and learning by using, possibilities for second-order learning. The case of car-sharing demonstrates that it is important for developers to examine their assumptions about users' needs and preferences and that it is equally important for users to have an environment that encourages them to question their needs and preferences.

The Praxitèle experiment, like the cases discussed in Chapter 4, is an example of the observation that users are not inherently innovative. Users were interviewed, followed, and incorporated in the Praxitèle experiment, but no second-order learning has taken place. Users tested and learned how the public cars could augment their mobility, but they showed little change in their mobility patterns, nor did they begin to question those patterns.

Nevertheless, after a slow start due to funding problems, Praxitèle has been a successful experiment. It resulted in many important findings about the complex set of technologies involved, about possible user acceptance, and about the scheme's relationship to conventional public transport. These results provide a good basis for larger-scale follow up activities that are being planned in other cities. Praxitèle has also shown that combining electrified mobility with reconfiguring mobility can have promising results.

Lessons for strategic niche development

Bikeabout, ASTI, and Praxitèle were planned, top-down organized experiments, while car-sharing was a newly emerging experimental business. All four were successful in terms of the general objectives defined at the outset of the experiments. In many respects, all four reflect good management practice. Most of the goals were realistic in the context of the experiments, yet also challenging enough to enable a broad range of learning processes and to gain commitment from the partners involved. Necessary networks were formed, market analyses were carried out, and the technologies involved were carefully selected.

However, a major difference between the car-sharing and the other three projects is the resulting market niche development. Car-sharing has created a bandwagon effect and has evolved into an expanding market niche. The other three are part of a process of niche branching, with technological and market niches involved, that, due partly to the experiments themselves, is promising but still in its infancy.

How can we explain this difference? Must we conclude that formally organized experiments have less potential for Strategic Niche Management? This would be a superficial conclusion. The Swiss car-sharing project had many features of an experiment, and we would like to single out two factors that were successfully dealt with by the actors involved. In fact, these factors are the two dilemmas first identified in Chapter 4.

The first is the contradictory set of pressures and trade-offs relating to achieving success in the short run versus reaching long-term goals. The actors involved in car-sharing have built their experiment, and their business, through a number of phases. In each phase, they set different goals that were not predetermined, but emerged partly from the process. Therefore, learning was a front and centre part of the process. The important lesson is that reconciling short-term and long-run objectives can be achieved by taking enough time to experiment, but doing so in a modular (or phased) fashion. Building new markets is a long-term process that cannot be organized overnight. Trying to achieve goals that are too challenging in too short a time will create disappointment. Government funding might be a mixed blessing in this regard, as the actors involved in the car-sharing project realized. Such funding could result in pressures to force the experiment forward too quickly and, ironically, give the project too much protection. This could result in a 'dinosaur effect', where some of the new practices and technologies are incubated in niches that cannot survive without continuing external protection.

The second dilemma involves the trade-off between the need, on one hand, to involve conservative actors who may actually prefer to develop the dominant regime but have resources and the power to mobilize, and, on the other hand, the need to keep the innovative spirit alive. It is difficult to build a broad network with highly aligned actors. In the car-sharing case, such a network was able to emerge by building on the strong involvement of users who provided resources and legitimacy, and on the work of intermediary actors such as the VCS traffic club.

Car-sharing was also the only experiment that resulted in substantial second-order learning. Mobility patterns were questioned and re-examined. For this reason, car-sharing has become a most promising avenue for a regime-shift. The literature of innovation has claimed and shown that high user involvement is a necessary condition for radical innovation (see Chapter 2). The car-sharing case confirms this finding, but we may make the conclusion more specific. For example, user involvement in Praxitèle was also high, but the nature of the involvement was quite different. In Praxitèle, users were the targets of analyses and questionnaires, while in the car-sharing case users were designing the experiment themselves. A context was created in which users were free to set their own agenda and therefore ask more fundamental questions about reconfiguring mobility.

Reconfiguring mobility revisited[58]

A Business Day Trip: Bristol to Oxford – the Present and the Future

Suppose a businessman in Bristol needs to plan a day trip to Oxford, 120 km away.

The Present . . .   Time

Check oil and tyres of car the night before

The planned route will use first a local motorway (10 km) which links with the main East-West motorway. After following this road east for 60 km, the route uses a local road for the remaining 50 km.

Prepare to leave home	6.45
Find car keys and get into car	6.50
Drive to local petrol station	6.55
Fill petrol tank and buy snack for breakfast; pay by bankcard	7.00
Drive through city commuter traffic	
Encounter road works in city centre	
Join congested traffic on motorway	7.15
Drive 120 km to Oxford	

Drive at an average speed of 90 km/h

Eat breakfast in car while driving . . . feel tired . . .

Listen to radio during journey

Arrive on the outskirts of Oxford City	8.35
As parking is full in centre, signs direct you to a park-and-ride	
Find parking space, look for change and pay with cash	8.40
Wait for bus to town	
Bus arrives for town centre	8.50
Bus takes you to central stop	9.00
Walk remaining 0.5 km to office . . . arrive a little late . . .	9.05

Need to be in Oxford office by 9.00 am

The Future . . .   Time

Log onto the Internet the night before. Order a dial-a-ride taxi and a MonoPod* for the next day's journey. Pay for the whole trip with smartcard.

*'MonoPods' use an upgraded high-technology national railway track network. They are autonomous rail modules that travel either independently or link together to form 'MonoTrains'. All pod transits are autonomous although they are co-ordinated by a central computer.

Dial-a-ride taxi, a solar-assisted hybrid minibus, arrives and takes	7.00
businessman to the local MonoPort via reserved vehicle lanes	
Arrive at the MonoPort	7.10
Check in at desk	7.15
Enter reserved personal Pod	7.20
MonoPod leaves Bristol. En route, the computer-controlled pod links	7.25
with other pods going in the same direction toward Oxford. During transit,	
work, rest, and watch TV. Request a stop at any port en route if necessary. Pods can	
travel at a maximum speed of 140 km/h. Eat a light breakfast in comfort. Thirty km from	
Oxford, some pods disengage and continue on separate tracks to London. Three	
Oxford-bound pods, continue on to the central Oxford spur.	
Arrive at Oxford MonoPort in the city centre	8.30
Pick up a one-seater Micro-EV from the rental stand; the day's rental is	8.35
included in the pod ticket, and the EVs also contain collapsible	
bicycles for local use.	
Drive to the office on reserved Micro-EV lanes	

| Park in Micro-EV-only parking | 8.45 |
| Enter the office at 8.50 am with ten minutes to spare | 8.50 |

. . . as it is a sunny day, decide to cycle down to the river at lunchtime on bicycle supplied with the EV . . .

Notes

1 Hård, M. and Knie, A. (1993) The ruler of the game: the defining power of the standard automobile, in *The Car and Its Environments. The Past, Present and Future of the Motorcar in Europe*. Proceedings from the COST A4 Workshop. Trondheim, Norway, *op.cit.*

2 As early as the 1930s it was an explicit goal of the city officials to make Los Angeles suitable for car traffic and to give up public transport infrastructure.

3 The case of automatic zone access systems in Bologna illustrates this quite impressively. See Hoogma, R. (1998) Introduction of Automated Zone Access Control in Bologna, Italy. A Case Study for the Project 'Strategic Niche Management as a Tool for Transition to a Sustainable Transportation System'. Enschede: University of Twente.

4 This case study is based on Black, C. (1998) The BIKEABOUT Experience. An automated smart-card operated bike pool scheme. A case study for the EC DG-XII supported project 'Strategic Niche Management as a Tool for Transition to a Sustainable Transportation System'. Milton Keynes: The Open University.

5 Potter, Steve (1997) *Vital Travel Statistics*. London: Landor.

6 Wood, C. (1996) Integrating Cycling and Public Transport. TransPlan Occasional Papers, No.1.

7 University of Portsmouth (1992) *A Mobility Policy*. Portsmouth: The University.

8 The reason was that the County Council had already funded a project in Southampton, the other large city in the county, and felt the obligation to redress the balance by funding a project in Portsmouth too. The ENTRANCE (ENergy Savings in TRANsport through Innovation in the Cities of Europe) project was supported by the Directorate-General for Energy (DG17) under the Thermie programme, the aim of which was improving the quality of energy utilization in the European Union and reducing its dependence on imported fuels.

9 The ENTRANCE money covered 40% of the funding for the scheme, providing that the county contributed the additional 60%. The county money was a loan which had to be approved by the Department of Transport, and this approval was given reluctantly because the county had been lent money shortly before for another European project in the area of transport. The total budget of Bikeabout was app. £200,000.

10 The users pay a deposit of £5 for the smartcard, and sign an agreement to pay up to £150 if they fail to return a bicycle.

11 The bicycle has to be returned within three hours. Penalty points can be added to a smartcard if the bicycle is returned late or is found to have been damaged. An automatic 'hold' can be put on the cards of people who abuse the system and the system administrator will be notified.

12 Initially, consultants hired to consider the viability of the Bikeabout concept reported negative comments from other cycle-scheme operators who suggested that 'It would be cheaper to buy 1000 specially painted bicycles with good insurance and heavy locks, with enclosed lock up for the night . . . It is not worth the cost to develop an electronic system. It is an expensive luxury to know where each bike is at a given moment.' TAS/Diepens and Okkema, Bikeabout (1994) Advanced Technology Cycle Parking Inventory. Consultant's report prepared for the ENTRANCE project, p. 23.

13 Black, C.S. (1996) Monitoring and evaluation of the ENTRANCE project: Portsmouth University Transport – Report of the 'before' surveys.Transport Research Laboratory/ University of Southampton/University of Portsmouth; Black, C.S. (1997) Monitoring and evaluation of the ENTRANCE project: Portsmouth University Transport – Report of the 'after' surveys. Transport Research Laboratory/University of Southampton/University of Portsmouth.

14 This case study is based on Potter, S. (1998) Greening Transport for Disabled People. A study of the Camden Community Transport's 'ASTI' project . A case study for the EC DG-XII supported project 'Strategic Niche Management as a Tool for Transition to a Sustainable Transportation System'. Milton Keynes: The Open University.

15 Passant, E. (1996) The Camden ASTI project, in *ASTI Technical Seminar*. Nuneaton: Motor Industry Research Association.

16 The consultant was Richard Armitage.

17 Oxfordshire County Council (1995) *Oxford Electric Bus Project – the first 500 days. An operational and environmental report*. Oxford: Oxfordshire County Council.

18 Green, A. and Charters, D. (1996) Camden ASTI project: electric and natural gas fuels for accessible transport, in *ASTI Technical Seminar*. Nuneaton: Motor Industry Research Association, p. 1.

19 The CCT co-ordinator, Ed Passant, had been active in debates over a number of strategic CT issues, and he was also an officer on the Executive of the national Community Transport Association (CTA).

20 PowerGen also had an interest in power control technology for electric vehicles; after the start of ASTI, PowerGen acquired a 50% holding in Wavedriver.

21 Harvey, V.J. (1996) Community transport for the year 2000, in *ASTI Technical Seminar*. Nuneaton: Motor Industry Research Association.

22 The delay led to inefficiencies in fleet utilization, as buses had to return to the depot for recharging whereas they could have been used for another service if a second recharging facility were available. The second recharging station was erected in 1997 in the south of Camden in addition to the station at the CCT depot in the north of the Borough.

23 Quoted in Claydon, K. (1996) ASTI vehicles – their use in the real world and user acceptance, in *ASTI Technical* Seminar. Nuneaton: Motor Industry Research Association.

24 Community Transport (1996) Major new CT project for London. *Community Transport*. Hyde, Cheshire: Community Transport Association, January, p. 5.

25 The car-sharing case study is based on Harms, S. and Truffer, B. (1998) The Emergence of a Nation-wide Car-sharing Co-operative in Switzerland. A case study for the EC DG-XII supported project 'Strategic Niche Management as a Tool for Transition to a Sustainable Transportation System'. Dübendorf: EAWAG.

26 Quotations in the car-sharing study refer to interviews, which were conducted by Harms and Truffer in the context of the mentioned SNM research project. Two dozen interviews were carried out with founders, managers and the surrounding network of both ShareCom and ATG. Furthermore, about 40 users were interviewed individually and in groups in 1997 and 1998. A more detailed analysis is given in Harms and Truffer, *op. cit.*

27 Both initiatives independently chose to organize as a co-operative. In Switzerland, a co-operative is very easy to found and its legal status guarantees a high degree of participation. In other countries, such as Germany, different forms had been chosen, due to a different legal framework.

28 Muheim, P. (1998) *Car-sharing – Der Schlüssel zur kombinierten Mobilität. Synthese.* Bern: Bundesamt für Energie/Energie 2000.

29 See also Muheim, *op. cit.*

30 Statement of a ShareCom manager.

31 The system was invented and tested by ShareCom, and was now ameliorated.

32 VCS, one of three big traffic clubs in Switzerland, is explicitly promoting public means of transport, cycling and walking.

33 The developments in Germany followed quite similar lines. Here the VCD (*Verkehrsclub Deutschland*) played the role as a mediator. For further details see Harms and Truffer, *op. cit.*

34 The situation in Switzerland thus differs from the situation in the Netherlands, where the car-renting firms are among the big providers of what they call 'car-sharing'. See Harms and Truffer, *op. cit.*, and Meijkamp, *op. cit.*

35 See Wendt-Reese, S. (1997) Der Dachverband European Carsharing/ecs e.V., in Bildungswerk

Weiterdenken und Birger Holm (ed.) *Carsharing im Vergleich. Das Autoteilen als Verkehrsträger innerhalb des Mobilitätsverbundes.* Dokumentation zur Fachtagung 'Carsharing – ein ökologisches und verkehrspolitisches Modell für ostdeutsche Grossstädte?' Dresden.

36 Statements of ATG and ShareCom founders in the interviews.

37 See also Kemp, René, Truffer, Bernhard and Harms, Sylvia (2000) Strategic Niche Management for sustainable mobility, in Rennings, K., Hohmeier, O and Ottinger, R.R. (eds.) *Social Costs and Sustainable Mobility – Strategies and Experiences in Europe and the United States.* Heidelberg: Physica Verlag (Springer), as well as Truffer, B. and Kemp, R. (1998) The Social Construction of a New Mobility Form – Experiments with Organized Car Sharing. Paper presented at the EASST conference, Lisbon.

38 This role was especially important in the beginnings of the process. With time, however, the contacts could be established directly without any major problem.

39 Statement of a user of the *ShareCom* system.

40 See Aarts, H. (1996) Habit and Decision Making. The case of Travel Mode Choice. Dissertation Katholieke Universiteit, Nijmegen. For a more detailed analysis of the motivations to join a car-sharing organization, see Harms, S. and Truffer, B. (2000) The Long Way from Interest to Participation: When Does the Car Owner Change to Car Sharing? Paper presented at the 79th Annual Meeting of the American Transportation Research Board, January 9-13, Washington, DC.

41 By using the term 'natural', the interviewee wanted to stress that he began to see his former behaviour as somewhat problematic and non-natural.

42 Muheim, *op. cit.*

43 See Harms and Truffer (2000) *op.cit.* for a more elaborate treatment of this problem. See also Franke, S. (1997) Gewohnheiten und ihre Veränderungen im individuellen Mobilitätsverhalten. Internal Manuscript. Zentrum Technik und Gesellschaft, Technische Universität Berlin.

44 See also MVV(1996) *MVV und Car-sharing. Ergebnisse einer Repräsentativ-Befragung von Kunden der Münchener Car-Sharing-Organization 'STATTAUTO'.* München: Münchener Verkehrs- und Tarifverbund.

45 See Energie 2000 (1997) *Facts & Figures.* Zürich: Ressort Treibstoffe.

46 The management of ShareCom and ATG stated that *'in the last ten years the media made free publicity for car-sharing equivalent to several million Swiss Francs'.* Most new members indeed got to know about car-sharing via the media or by word of mouth (Harms and Truffer, 1998, *op. cit.*).

47 This case study is based on Simon, B. (1998) The Praxitèle Experiment of Self-service Rented Electric Vehicles. A case study for the EC DG-XII supported project 'Strategic Niche Management as a Tool for Transition to a Sustainable Transportation System. Maastricht: MERIT.

48 Bénézra, C. (1994) Les véhicules électriques en libre service: politique des villes et stratégies des acteurs. Thesis, Institut des Études Politiques de Grenoble.

49 Faudry, D. (1997) Véhicules individuels publics et auto partagée. *IEPE*, February.

50 CGFTE stands for Compagnie Générale Française des Transports et d'Exploitation.

51 INRIA stands for Institut National de Recherche de l'Informatique et d'Automatisme; INRETS is the acronym of Institut National de Recherche sur les Transports et leur Securité.

52 Parent, M. and Texier, P-Y. (1993) A Public Transport System based on Light Electric Cars. Paper presented to the Fourth International Conference on Automated People Movers, Irving, Texas, March.

53 Massot, M.H., Broqua, F., Polacchini, A., Blosseville, J.M. and Dumontet, F. (1997) Prior Technical and Economical Evaluation of the Station Car System Praxitèle. Paper presented to the ITS Congress, Berlin.

54 Based on information leaflets available from the Praxitèle consortium.

55 CGEA stands for Compagnie Générale des Eaux. This large multinational corporation with operations in many branches of industry has meanwhile changed its name to Vivendi.

56 These and following statistical results are taken from Carli, A. (2000) *UTOPIA Test Site*

Report Praxitèle Saint-Quentin en Yvelines. Report for the UTOPIA project. An intermediate report is Bleijs. C.A. *et al.* (1998) Results of the Experimental Self-service Electric Car System Praxitèle, Proceedings of the 15th Electric Vehicle Symposium, Brussels, October.

57   Urban transport in the age of information. *Renault R&D*, No. 15, January, 2000, pp. 54–56.

58   By Ben Lane, taken from Weber, M., Hoogma, R., Lane, B. and Schot. J. (1999) *Experimenting with Sustainable Transport Innovations. A Workbook for Strategic Niche Management.* Enschede/Sevilla: University of Twente.

Chapter 6

Strategic Niche Management

Innovation studies have made a lot of progress over the last decades in understanding technical change. They have showed that technical change is not a haphazard process but typically patterned and highly cumulative. In the havoc of the world, paths emerge, through the use of technologies, giving rise to path dependencies for which no one has really opted. These factors act as constraining and shaping forces upon the choices of actors involved in technical change. In Chapter 2 we introduced the concept of technological regime to refer to the correlated set of factors that gives rise to rules to which the actors adhere consciously and unconsciously.

At the same time, well-trodden paths are left behind, new paths are created and existing ones meander in unexpected ways. Technical change is not a linear process, but rather erratic, in which chance events are operating in tandem with structural factors. This raises the question how new paths are created? Do people create them: system builders, men with imagination and diligence, who manage the junctions of multiple market places and political intersections? Or are they created by the anonymous operation of supply and demand in a landscape of changing values, beliefs, politics, rising income, growing scarcities and so on, and a changed physical environment?

In this book we come up with a different explanation for how paths are created and regimes are transformed: through the emergence of novelties in so-called niches that offer a resource base for further development. Niches are domains of application with distinct selection criteria that in some respects differ from those defined by the technological regime. The central assumption of this book is that niche development processes play a crucial role in this process of breaking path-dependencies and creating new paths. We offer a new perspective, much more than a new explanation. The niche development mechanism is not something that operates separately from the two other mechanisms. In niche development there is governance (agency) and there are changed cost and demand conditions, but these are viewed from a different perspective: a multilevel perspective of socio-technical change – of niches, regimes and the landscape, each with their own dynamics and influence.

The creation of paths is an important issue for sustainable development because sustainable development requires fundamental change. This book contends that there are no technical fixes to the problem of sustainability.

Behavioural changes are needed too. There is not only the erosion of natural capital for which some technical fix may exist but also the erosion of social capital, through consumerism as a way of life and mind-numbing jobs, things for which there are simply no technical fixes. We believe that each technical fix will create its own 'nightmare', which is hopefully less harrowing than the one we are experiencing today (of urban and global pollution, noise, road accidents and congestion in the case of car-based transport). What we need is technical change *and* social change. And this book suggests a way to achieve this, which is Strategic Niche Management (SNM). Strategic Niche Management consists of the experimentation with new technologies in society, but also includes policies to diffuse further technologies that help to create a regime transformation and research programmes for pathway technologies. Here the focus is on technology experimentation involving real users.

The message of this book is one of hope, or, more precisely, of moderate optimism. It is based on the belief that SNM may be used in creating new paths and for working towards structural change that produces sustainability benefits. Yet, a clear perspective on how to do SNM and manage niche development processes is missing. Our goal has been to fill this gap and develop a perspective on how to nurture processes of niche formation leading to sustainable development, and in this context SNM refers to the process of deliberately managing niche formation processes through real-life experiments. The core idea is that through experiments with new technologies and new socio-technical arrangements processes of co-evolution can be stimulated. Technologies – for example electric vehicles or smartcards – as well as the contexts (user preferences, networks, regulation, complementary technologies, expectations) in which they develop are worked upon simultaneously. In other terms, SNM aims at *aligning* the technical and the social. As a consequence new, more sustainable mobility patterns might emerge, partly embodied in hardware (new technologies) and in new practices based on new experiences and ideas.

The need for a shift towards sustainable technological regimes is clearly defined in policy arenas, nationally and internationally. This book contributes to formulating such policies. The analysis does not lead to simple recipes for what to do. In contrast to much of the literature on environmental protection and sustainable development that focuses on the best instrument to achieve set goals, we do not see policy-making as an instrumental activity. It can best be defined as an interaction process involving differences of interests and views, conflicts, negotiations and coalition building among actors. The approach of SNM we have developed offers a new perspective on the introduction of new technologies and helps to identify a number of useful actions to be taken. Most of all it leads to enhanced reflexivity: actors will get a better understanding of the nature of technical change and dilemmas in managing processes of change. Complexities and dilemmas are not shunned but brought to the fore and discussed.

This Chapter will bring together the lessons from the experiments as seen through the lens of Strategic Niche Management and the lessons for Strategic Niche Management as a perspective on innovation. The focus of the book is on societal experimentation with new technology, focusing on two main issues: *what has been learned* (and not been learned) thanks to the experiment, and to what extent did the experiment lead to *institutional embedding* in the form of the creation of new networks, public acceptance, and strategies and actions of private and public actors to further the wider use and development of the technology experimented with. We looked at learning and spinoffs, which we related to the set-up of the experiment. We did not analyse processes of niche formation as such,[1] let alone the process of regime-shifts that could benefit from successful niche formation.[2] However, the material collected allows us to draw some conclusions on these issues as well.

In Chapters 2 and 3 we defined Strategic Niche Management and talked about lock-in and the role of niches in regime-shifts. These Chapters can best be read as a series of explorations of the idea of SNM, which set the stage for a description of eight societal experiments with more sustainable vehicles, car use, individualized public transport and non-motorized forms of transport, such as bicycle use from a bicycle pool. The eight experiments are positioned within two development lines towards more sustainable transport: electrifying mobility and reconfiguring mobility. In Chapter 4 we described and discussed the four experiments with electric propulsion and in Chapter 5 the four experiments with reconfiguring mobility. We have only highlighted those aspects that help us to improve and develop the idea of SNM. This means that the sample is biased and not representative for experiments with transport innovations offering sustainability benefits. Each case is unique and there are many ways of aggregating them. We have aggregated them on the basis of the above distinction (of electrifying mobility and reconfiguring mobility) and the type of learning that occurred as a result of the experiment. In this final Chapter, we will present a comparative analysis of these eight experiments. We will start this Chapter with a summary of the various contributions the eight experiments studied made to processes of niche development.

Experiments and niche development

Rügen – an experiment of vehicle testing which failed to investigate the conditions for alternative vehicle use

The Rügen project was a large-scale experiment in Germany with 60 prototype battery-powered electric vehicles on the Island of Rügen. The vehicles were converted internal combustion engine vehicles produced by German automobile manufacturers, which were being used in turns by 100 users over a period of four years.

The experiment did not contribute to niche development for battery driven electric vehicles – in Germany or anywhere. The niche created was a temporary technological one which did not flourish afterwards. Networks developed over the course of the experiment were limited in scope as they consisted mainly of producers of cars and components. Broader networks that could have carried further developments did not emerge and no EV infrastructure was developed. The main outcome of the experiment was a confirmation of the German car industry's negative expectations about battery EVs. This helped legitimize their choices for other alternatives such as fuel cells.

Learning consisted mostly of learning about technical aspects such as the battery, electric drive, fast charging and environmental impacts. This was no coincidence as the experiment was explicitly set up towards these ends. There was very little second-order learning about users and new types of mobility. The project did not study the conditions for EV use: the initial assumption that businesses and fleets would be the users was not explored, but simply assumed and then confirmed. Yet the project did reveal one unintended interesting finding about EV use, which was that users (for a short period) used it as their first car.

PIVCO – a new car that fell prison to its context

PIVCO was an outsider initiative in the niche for electric vehicles. The car was designed and built by Norwegian companies with no background in car production. The aim of the project was to develop and produce an electric vehicle for urban and suburban transportation, thus for private consumers. The vehicle, called City Bee, is a small two-seater EV built on a spaceframe of welded aluminium with a thermoplastic body. Unlike the vehicles tested in Rügen, the City Bee is a specially designed vehicle around a battery propulsion system.

Throughout the experiment a great deal was learned about user acceptance. User acceptance was high in Norway but low in California where the vehicle did not meet the users' demands. One reason why the vehicle was not a success in California is that the City Bee's top speed was too low to be used on the highways. It seemed that the niche for the City Bee would be limited to Norway where the vehicle capitalizes on national sentiments and a range of support measures. The experiment revealed several design faults, which led to subsequent improvements in the four generations built. There was no co-evolution of regulatory and customer demands, so that the learning just led to a redesign of the vehicle to fit existing demands better.

Unlike on Rügen, to some extent the PIVCO experiment did lead to societal embedding. Production facilities were created, recharging stations were built and a maintenance network was set up. The carrying network was strong, but limited as neither consumers nor government agencies were represented. Expectations did not develop much over the course of the experiment; they hardly became more widely shared or more specified. Despite all efforts, the technological niche did not develop into a market niche, and the company suffered losses. It entered safer waters when it was bought by the US carmaker Ford, who intends to turn the City Bee (now called TH!NK) into a global product for market niches for both fleet and private cars.

This case shows that technological niches can become like prisons, hard to escape. The PIVCO experiment created a small potential market in Norway. But this market was the result of distinct and specially created circumstances in Norway that cannot easily be copied in other countries. The special circumstances led to niche creation, which appeared to become a dead-end, had not Ford stepped in and added multinational corporate protection to these special protective circumstances.

La Rochelle – a case of misreading the market

From late 1993 to late 1995, 50 electric vehicles were driven by people using them to meet their transport needs in the French town La Rochelle. The project was a market-research one, aimed at testing the market for electric vehicles. Private and business people used 25 prototype Peugeot 106s and 25 prototype Citroën AXs in this period at a subsidized leasing fee. The experiment arose from the existing co-operation between the three main actors: PSA, EDF and the Municipality of La Rochelle. A technological niche for EVs already existed before the experiment, created under the guidance of EDF, the national electricity producer which had been using EVs within its fleet since the 1970s. The aim of the experiment, however, was to investigate the household market. The main actor, PSA was convinced that EVs were attractive for households and wanted to demonstrate that the technology was ready. As in Rügen, the vehicles used were converted ICEVs.

Although experiences from users were very favourable, prompting PSA to invest in the commercialization of EVs, there was merely a gestation of technological niches. Actual demand was much lower than expected. If we look at what has been learned we see important lessons about users' experiences and behaviour. Some learning on technical aspects also took place (with regard to the cable and reverse button) which allowed companies to improve the product. Societal embedding has been limited. As noted, the project led the automobile producer (PSA) to invest in small-scale production but sales were disappointing. The sales projection was 2,000 cars per year rising to 10,000 (full capacity of the plan) by the year 2000. In fact PSA sold little over a thousand EVs per year in the period 1995-2000. In other words, expectations about market success became stronger but the market failed to develop.

The La Rochelle experiment showed that under the current regime of car use, EVs for private use cannot survive without special protection. The technology could not escape the phase of a technological niche, but it was important that developments with EVs in France did not come to a halt. The experiment was followed by other experiments in La Rochelle, including the public use of EVs (Autoplus and Liselec). Several public transport companies have become interested in the use of electric vehicles in their fleets. The niche for EVs is still alive in France but not flourishing. The announcement of the introduction of range extenders may increase the market chances for electric vehicles.

Mendrisio – the idea of SNM put into use (but not into currency)

In the Swiss town of Mendrisio, a large-scale experiment was conducted with lightweight electric vehicles (LEVs). The main goal was to substitute a proportion of 8% (i.e., about 350) of all vehicles by electric vehicles – not just automobiles but also lightweight vehicles and motorcycles. This substitution should be achieved by a whole series of promotional measures.

The six-year experiment was initiated as part of a federal promotion programme for LEVs in 1995 in order to simulate the market diffusion of the technology. This approach was motivated by the belief that the main problem of the current LEV-technology was not so much the technological weakness of the vehicles, but rather achieving of economies of scale at the level of manufacturing, infrastructure and the market. Before the experiment, about 2,000 LEVs had been sold in Switzerland. For several years the market for LEVs had experienced exponential growth rates, but by 1992 sales numbers began to fall and reached a constant level of about 100–200 vehicles per year. The Mendrisio experiment can be seen as a means actively to support the market niche of LEVs.

The project made a notable contribution to niche development in Switzerland. The project involved learning beyond the conventional technical testing of vehicles and market assessment of potential sales. Assessment of promotional measures took place and lessons were pursued on how to scale-up the use of LEVs in Switzerland and beyond. Peoples' mobility patterns were also assessed; one issue explored was the question of whether LEVs can constitute an element of new, integrated forms of individual mobility. It may be too early to draw definitive conclusions, but analyses of mobility patterns of LEV-users in Switzerland as well as other European countries show that

LEVs do have a potential for leading to a more sustainable individual transport system.[3] LEV users undergo a number of qualitative learning processes. Their personal modal split changes (the LEV becomes the most frequently used vehicle), they drive more cautiously and become more conscious about energy use in transportation. On the manufacturing side, there were no dramatic cost reductions through scale economies. The scale of production was too low for that, and learning economies take time. The sustainability dimension became weaker during the project.

Of the four experiments with electric vehicles, the Mendrisio experiment contributed the most to niche development, by creating a positive image for electric vehicles and a constituency and set of support measures for its wider diffusion.

Bikeabout – a shared bicycle scheme used by few automobile users

The Bikeabout experiment in Portsmouth, UK, is a bicycle pool experiment, involving the use of an electronic access and billing system, to address the weaknesses identified in earlier bicycle pool projects, notably bicycles getting lost or stolen. It applied advanced technology to provide a controlled and protected environment for the shared bicycle scheme. The experiment consists of three depots with individual racks that are accessible only with smartcards issued to registered users, and CCT surveillance gives additional security. Users can use the smartcards also to make reservations. The University held the experiment with the support of the local council. The University wanted to offer an alternative for the many car trips made between its two sites, which are only 3 km apart. The City Council provided a safe cycle route between the depots at both sites.

The project was successful in solving the problems of earlier cycle pool schemes. Staff and students started off with negative expectations of such schemes, feeling that they generally did not work because bicycles would be stolen and maintenance would be poor. The project changed this as it has been shown to be reliable; it worked well and no bikes were stolen or vandalized. It was recognized that the high-tech solution worked, but criticism also arose of the high costs. Moreover, Bikeabout did not change the attitude of people towards cycling and achieved only a marginal reduction in car use between the two sites. The project did not facilitate any major change in people's travel behaviour, as there were no complementary measures (particularly disincentives to car use). The lack of an integrated package of measures was an obstacle to second-order learning.

The experiment did not contribute much to societal embedding of the cycle pool. The project was designed to solve a local problem, but was not intended as a pilot for wider implementation of the electronic bicycle standards and reservation system. The network carrying the experiment was limited in composition. The official project partners, the University and the City Council, did not foresee commercialization of the bicycle pool technology, but in the course of the experiment the bicycle rack supplier became interested and took over the rights for marketing the concept. This increased the chances for niche development, but the supplier in fact only found one other client in the city of Rotterdam. The promising technology for protected cycle pool depots thus basically remained in a technological niche.

ASTI – more than a bus experiment

The ASTI (Accessible Sustainable Transport Integration) project involved the development and introduction into service of three electric and three CNG-powered minibuses accessible to all people with reduced mobility in the London Borough of Camden. Part of the experiment was the introduction of demand-responsive servicing using telematics technologies for vehicle positioning and route guidance, and scheduling software to optimize the use of these and other minibuses. The project was integrated into the development of an operators' network, which was to allow the pooled use of the fleets and better distribution of resources. The ultimate aim of the ASTI project was thus not simply to develop a small fleet of electric and CNG minibuses, but to develop a better and more efficient community transport (CT) service.

The project had a strong combination of partners for whom ASTI was important for developing their core activities and individuals managing the project who for their own reasons wished to 'champion' it. A prominent feature of the project is the co-operative culture of the initiating community transport organization, which together with its counterparts elsewhere in the country has always worked in partnership with its suppliers (their vehicles are specialist conversions), with public authorities (from whom they obtain financial support) and users (who elect CT management committees). This experience made Camden CT a suitable organization for a complex demonstration project with multiple stakeholders such as ASTI. The main actor absent from the project was London Transport, the mainstream bus operator, but it joined the follow-on project.

The ASTI project showed the viability of CNG and electric bus technologies for community transport applications. While CNG is already becoming established in the UK in other applications, costs of the latter remain a major barrier for adoption. The contribution of the project to niche development for those two technologies was therefore limited. More effective was the demonstration of the software systems and the concept and economics of integrating services via information technology. Besides applications to co-ordinate parallel fleets of vehicles, the vehicle location

(GPS) and scheduling technologies are very transferable and are of interest to individual fleet operators such as mainstream bus operators, taxi companies, dial-a-ride services and ambulance operators. Community transport has proved to be a fruitful niche for developing and demonstrating new technologies for transport.

Remarkably ASTI tends to be viewed by other transport organizations that run accessible services and even some of the partners, as a *bus* technology development project rather than a system development project. The project's basic vision of demand-responsive services for the mobility impaired was not shared and perhaps not even grasped. The concept was not taken up by these companies but was successfully applied in another London borough by CCT and Ealing CT in a joint competitive bid. CCT is now developing further the idea of a control centre through the Plusbus interactive scheme. The scheme consists of the pooling of resources and aims to improve integration and utilization of transport fleets from different operators.

Car-sharing – an unexpected success story of SNM

Organized car-sharing can be considered as a service innovation, which involves a new way of making cars available to users. People borrow a car from an organization of which they are a member. They pay on a per-vehicle and per-km basis. The first country that developed organized forms of car-sharing was Switzerland, but not much later, from 1988 onwards, and independently from the Swiss events, car-sharing was also introduced in Germany, Austria and the Netherlands. Today there are some 80,000 members of car-sharing organizations in the four countries, and annual growth rates are relatively high. In Switzerland 35,000 people are members of Mobility, the nation's largest car-sharing organization.

It was the existence and diffusion of new information and telecommunication technologies that enabled the success of car-sharing, as it made it easy to reserve a car. Cars are accessed by a smartcard, which is also used for administrative processing. In order to reach a broad market, the support of different societal actors was essential. Besides official promotion programmes launched by national agencies, traffic clubs, environmental organizations, the media and others also helped the car-sharing organizations on their way.

Initially the protection came from users who provided cheap money and took care of small repairs and the cleaning of the vehicles. During the growth of car-sharing, the providers as well as users underwent a number of learning processes. Car-sharers who are former car owners reduce their car use and make use of public transport more frequently, while former non-car owners retain a low level of individual mobility. Providers learned that they can increase their customer base if they integrate car-sharing in multi-modal mobility services. These second-order learning processes seem to be inherent in the concept: the reservation system and cost structure of car-sharing organizations hinder consumers from making use of cars habitually.

The evolution of car-sharing is the most successful example of Strategic Niche Management among the cases studied. The initially unplanned bottom-up processes involved close interactions between car-sharing providers and users, which enabled second-order learning to take place. The broad support from various sides resulted in effective societal embedding; the concept of organized car-sharing is well-known and accepted. Further, there exists a fine-meshed network of car-sharing stations and there are also co-operative ties with the Federal railway company and several local transport companies. Expectations of the further diffusion of car-sharing were, and are, high, and have led to plans for expansion. Car-sharing is expected to become a key element for integrated mobility services and innovation thus proceeds beyond the initial idea of sharing cars and providing the facilitating technology. The niche is blossoming and may be the seed from which grows a new regime for passenger transport.

Praxitèle – a successful failure of individualized public transport

The self-service electric car rental system Praxitèle was designed to offer an environment-friendly, intelligent solution to urban transport problems, notably parking congestion and flexibility of public transport. It was positioned as a bridge between private cars and collective means of public transport, and in particular as a complement to the latter. The objective of the project in a French town was to confirm the feasibility of the system by testing it under real life conditions. The experiment started in late 1997 in Saint Quentin-en-Yvelines, a town in the suburbs of Paris. Fifty electrically converted Renault Clios were available from 15 strategically located stations. The goals were manifold. The project aimed to demonstrate the technical functioning, to investigate the public's acceptance of this new type of service, to find the optimal conditions for exploitation, and to gather the knowledge required for replication of the project on other sites. The economic feasibility of the self-service system was not central to the project.

The project yielded many lessons about these goals. The technology functioned well and the number of users was gradually growing; membership doubled from May 1998 to May 1999 from 400 to 800. The services were used in connection with public buses or trains. Mainly non-owners of cars rented the electric cars but a substantial share of the users (44%) did own a car and 29% were using cars from family members. For both groups the self-service cars formed a convenient addition to the spectrum of existing modes of transport, but they did not question their mobility patterns. Interestingly, contrary to expectations, the majority of users were not employees of local companies or tourists but residents of the area for whom the system offered a useful addition to the spectrum of transport options. Second-order learning hardly occurred, either on the side of the users or on the side of the service providers who merely found confirmation of their expectations about the working of the system. Contrary to the provider's expectation, however, companies were less eager to stimulate their employees to use the service especially because of its experimental and thus perceived temporary character.

The project thus created a lot of learning, which could benefit follow-on projects elsewhere, but the transfer of the lessons is not secured through the set-up of a transfer mechanism. The project also suffered from a weak internal network. Several of the partners sat back and were little more than suppliers, and the project did not develop an efficient communication policy because early difficulties caused by the complexity of the project made the partners aware that 'high winds blow on high hills'. Users played a role in the creation of the service as the idea was born of discussions with them, but large commercial firms developed the project further.

Nonetheless, the central actors now seem confident that there is a future for the self-service systems with electric cars and are continuing with their plans for scaled-up follow-on projects. In conclusion, therefore, the Praxitèle experiment did not launch the self-service system outside the initial technological niche, but it did potentially contribute to niche replication. The experiment did not pose a threat to the regime actors: the service provided is seen as a complement to collective public transport and is not believed to lead people to abandon their cars. The technological niche of self-service systems with electric cars is likely to become a market niche that may form a link between the regimes of car-based individual transport and collective public transport.

Improving experiments – the managerial lessons from SNM

Through the experiments a great deal was learned about the technical functioning of electric vehicles, scheduling software and electronic reservation and accessing systems for vehicles and about user satisfaction and behaviour. It was learned that electric vehicles did not hold great appeal for consumers as a substitute for traditional cars, but that lightweight vehicles (electric assisted bicycles, electric scooters and three-wheelers) are quite attractive, both from a user point of view and from a sustainability point of view. They promote silent and cautious driving and the careful planning of trips (the shortest route). Many things could not have been learned in another way, i.e. from surveys or controlled laboratory experiments. In some cases the experiments had unexpected effects. The use of electric vehicles boosted consciousness of energy consumption that led to reductions in energy use in households through energy conservation measures. The experiments also generated a great deal of attention from information media and specialists.

Through the experiments much was learned, about possibilities and shortcomings, but a great deal also was *not learned*. The experimental lessons could have had a broader range and wider impact, if the experiments had been set up in another way, with another partnership at another location (involving different users and transport situations), utilizing a design that was more flexible and adaptive. Most of the experiments suffered from weaknesses that prevented the actors from obtaining useful knowledge, especially concerning user needs and the conditions for alternative mobility. Whilst making mistakes is inherent in situations that involve uncertainty, some mistakes could have been anticipated and in our opinion prevented.[4] From an SNM point of view, the following design flaws were common in the experiments.

Insufficient user involvement

Users generally played a rather passive role in the experiments. They were hardly listened to, let alone involved in the set up of the experiment. It was assumed that their needs were known and fixed. But users can be a source of creativity and support, as shown in the experiments of organized car-sharing and also in the studies by Von Hippel (see Chapter 2). The most successful innovation in market terms – organized car-sharing – was a user initiative.

The actual involvement of users turned out to have a strong impact on user satisfaction and purchase decisions. The EV users in La Rochelle developed high satisfaction with the vehicles and half of them decided to buy the vehicle they had tested. PSA extrapolated this rate of satisfaction to the general market without realizing that the potential clients outside the experiment lacked the hands-on experience of driving the vehicles and accommodating them into their travel patterns. PSA thus overestimated the market potential and registered disappointing sales. This may only be rectified if mechanisms can be found to give potential users the same hands-on experience as the La Rochelle users enjoyed.

User involvement is not a question of getting representative users though, or getting specific groups such as pioneers, laggards etc. User involvement is about creating a platform for a number of users to experiment with new forms of mobility and technologies that could embody these new forms. Diffusion would then imply further network building with active contributions of users themselves, so it involves '*translation*',[5] not simple and mechanistic spreading of a specific artefact to a group with specific preferences and other characteristics.

Too much focus on technical learning

Most of the experiments consisted mainly of technology testing, with the technologies predefined. The starting point was not a local problem (of pollution, congestion, lack of accessibility, need for new types of economic activity), but a solution. As a result, little was learned about user needs and preferences, about the opportunities for users to change their behaviour, or about how users could meet their perceived mobility needs in more environmentally sustainable ways.

The predominance of first-order learning

The learning that occurred was mainly of the first-order type. Learning was about vehicles rather than about mobility and ways to achieve sustainable mobility. Users were not challenged to question their mobility needs. The experiments did not create enough room for processes of co-evolution to occur. Here we should add that co-evolutionary learning requires testing and implementation of a mix of technologies and policies, such as transferia, transport telematics, special bus lanes, bicycle lanes, changes in the tax systems, car-free zones, etc. Co-evolutionary learning does not occur through the use of single technologies.

Minimal involvement of outsiders

The experiments were mostly dominated by 'insiders' in the regime, who had an interest in the *status quo* and therefore did not go to great lengths to investigate alternative mobility solutions. Rügen provides the best example of this. The large German automobile manufacturers for whom the experiment was primarily a way of testing batteries dominated this experiment. They were not interested in the concept of electric mobility, but felt pressured to consider battery electric drive as an alternative technology for propulsion. They wanted to be prepared for the remote possibility that BEVs might become a market reality.

The projects were overly self-contained

The experiments were generally not linked to sustainability visions or to scenarios of niche development. Strategic considerations were important in the BEV experiments but they were limited to commercial issues. A more varied involvement of stakeholders could have broadened the perspective of actors involved in the experiment. Sustainability issues were only given concerted attention in the alternative mobility experiments, but even the sustainability dimension was weak and not articulated. The relationship between experiment and niche was not given much consideration or handled with reflexivity. Societal embedding issues, which are very important in niche management, were neglected.[6] Users could have been more actively involved, more empowered. Governments, who are an important actor for sustainability issues, could also have played a stronger role. The projects generally failed to build platforms for interaction and achieving societal embedding. There was too much of a technology push.

These issues are not meant as a criticism of the day-to-day management of the experiments as such. The experiments were not badly managed in an operational sense, within the framework of the set objectives. We do not want to levy a heavy criticism on the managers who ran the experiments but feel that the projects could have benefited a great deal from the insights of Strategic Niche Management.

The relationship between the set-up of an experiment and its outcomes

In analysing the experiments we revealed an interesting correlation between the way in which the experiments were set up, and the outcome in terms of niche development. This may seem surprising given the importance of external factors, not under the control of the actors, in the process of niche development. We discovered that experiments with a high technical focus and limited orientation towards societal embedding (Rügen, but also La Rochelle and Bikeabout) failed to achieve second-order learning and co-evolution. They failed to escape the technological niche, to become a market niche or an element in a new regime.

The conclusion is that radical technologies require a special support effort with a societal embedding component. This fits with our hypothesis stated in Chapter 2 that radical technologies require a new regime whose emergence in turn depends on the occurrence of processes of co-evolution. The cases led to a further refinement of our hypotheses concerning the relationship between the experimental set up of the cases and outcome in terms of type of learning (first order versus second order). The refined hypotheses are given in table 6.1 in bold characters.

The table shows that second-order learning does not result in radical change unless there is a broad network involving outsiders. None of the EV experiments did well on second-order learning and societal embedding, *and* contributed to a regime-shift. The only case that did combine these elements is organized car-sharing, which involved both second-order learning and societal embedding, and has the potential to make a strong contribution to a regime-shift.

The argument for doing experiments with a potential for regime transformation should not be read as an argument against doing experiments that merely seem to preserve the existing regime. It is entirely possible that technical experiments run by traditional actors with no intention to transform

Table 6.1 Outcomes of experiments studied in terms of learning and societal embedding and hypotheses about the relationship between type of learning and network and niche development

Learning Carrying network	Low First-order learning (mostly technical)	High Second-order learning
Network dominated by traditional actors	**Exit or option stays in technological niche**	**Option stays in technological niche, or becomes element in existing regime**
	Rügen	Mendrisio ASTI Praxitèle
Broad network involving users and outsiders	**Option becomes element in existing regime or market niche develops**	**Option becomes market niche and/or becomes element in a new regime**
	La Rochelle PIVCO Bikeabout	Car-sharing

the existing regime will contribute to the development of a new regime. The experiments with battery electric vehicles in the 1980s which were done in a conservative way – by equipping existing cars with batteries rather than developing dedicated electric vehicles for specialized use (urban trips or to go to a transfer station) – may ultimately contribute to intermodal travel and electric mobility. Conservative experiments may thus contribute to a transition and the case for and against such experiments is not as clear as it appears to be.

A reverse logic may apply to radical innovations (such as lightweight electric vehicles and dial-a-ride services) that may end up being used in a conservative way (leading to more motorized transport). In fact, this happened in the large-scale experiment with electric mobility in Mendrisio. Innovations are not inherently disruptive or conservative, it is the way in which they are utilized, together with their technology features, that makes them either disruptive or conservative. Impacts are *co-produced* by technology and institutional arrangements.[7] What we need is experiments and experimental designs that help us to discover ways in which the two dimensions – technical design and social and organizational arrangements – work in harmony towards the goal of sustainable mobility.

What do the cases tell us about SNM?

After criticizing the experiments we turn to the question: what do the cases tell us about Strategic Niche Management? Do they confirm the assumptions about SNM stated in Chapter 2? Looking back, we see that some assumptions were not confirmed. For one thing, we were certainly over-optimistic about the potential of SNM as a tool for transition. There is an incongruity between what we expected and what we observed in the following respects.

The positive circles of feedback by which a technology comes into its own and escapes a technological niche are far weaker than expected and appear to take longer than expected (5 years or more). There are lead times in decision-making and existing ownership structures that act very much as an inhibiting factor. The experiments did not make actors change their strategies and invest in the further major development of a technology; efficiency improvements resulting in cost reductions were small (due to the small scales of production); and apart from organized car-sharing there has not been a takeoff or significant scaling up.

The contributions of the projects to niche development appear to be small. Much has been learned about the functioning of technologies and their acceptance. In all cases, new product constituencies have been built, some of which are quite powerful, such as the constituency for BEVs in France and the one for organized car-sharing in Switzerland. The constituency for Rügen was dissolved and the one for Bikeabout consisted of just two organizations. Niche development was slow or even absent, as in the case of Rügen. In some cases the market potential of a new technology or concept could be established thanks to the experiment, such as the ASTI fleet management software, the protective bicycle racks, and electric bicycles and motorcycles. In other cases, the experiments showed that there was not a real market, as for BEVs for private, individual use. Even the contribution to local problems was generally small: no spectacular results have been achieved here as the experiments were of too small scale. The experiments were relatively isolated events. It seems difficult for the actors to build bridges.

Although perhaps more could have been done and achieved there are limits to the power of experiments. Only occasionally will an experiment be such a big success that it will influence strategic decisions. Experiments may tip the balance of decision-making, as has happened in many cases, but they will not change the world in a direct, visible way. One should therefore not evaluate an experiment on the basis of its immediate economic success but on the basis of what has been learned and its contribution to processes of social embedding. Negative learning then becomes something useful. One also should accept that failures are inevitable in experiments and that it takes time for some of the lessons to be applied in ways that produce real benefits (either economic benefits or sustainability benefits). It took the Swiss car-sharing co-operatives 7 years before they achieved a membership of 5,000 people. We ourselves are convinced that the knowledge about electric propulsion and competences in lightweight, plastic-bodied vehicles obtained in the experiments with electric vehicles will be utilized in the future in hybrid electric vehicles and in internal combustion engine vehicles

Failure should be accepted in the quest for success. A good example of this is the Praxitèle experiment with self-driven public electric vehicles in St Quentin-en-Yvelines in France which was terminated because it was not economically viable, but the public transport company CGEA is convinced that the system has economic potential as a supplement to existing public transport services and will introduce it in future in various cities in France. It is often stated that the road to success is paved with failures and that for progress one needs deviants, two popular truisms with which we fully concur. Fortunately, not all experiments were failures in their own set objectives, and the case for experimentation does not have to be forcefully made by us or anyone else, as many people are convinced that societal experimentation with technology is useful.[8]

It is important that the experiments are seen as learning exercises and projects that are occurring in the *predevelopment* phase of a transition or technological regime-shift. In a predevelopment phase there is a great deal of experimentation with novelties in the form of trials, but no visible trail. This may lead people to conclude that change will come from other events and directions. But this would be a wrong conclusion. To get a feel of the future one should look at people's expectations, not at the results of experiments. There is a convergent view about the contours (but not the details) of future passenger transport, a view that is guiding strategies and investment by private and public decision-makers. All car manufacturers are convinced that alternative fuel vehicles will be on the roads in the next 5 to 10 years, some even think that in the future people will no longer own a car. Bill Ford, chairman of Ford, the world's second largest carmaker, has said that fuel cells will end the 100-year reign of the ICE and that we may witness an end to car ownership as the preferred method of personal transportation. In his words:

> The day will come when the notion of car ownership becomes antiquated. If you live in a city you don't need to own a car.

Under his vision, Ford and other carmakers could own vehicles and make them available to fee-paying motorists when they need access to transportation.[9]

It is unclear to what extent the rise of organized car-sharing informed this vision, but it is likely to have played some kind of part. Experiments influence the world (by changing the odds of certain futures) but do not bring particular futures about. Their influence is more indirect. It is only during the takeoff phase and especially the acceleration phase of a transition that we start to see the effects of explorations and niche projects.[10] This demonstrates the importance of doing a *wide set* of explorations and of designing experiments in ways that allow for second-order learning (about basic assumptions), besides first-order learning. As a last remark, regime transformations are associated not only with learning and adaptation but also with unlearning and undoing. The latter aspect (of dissociating old ideas, assumptions and habits) proves to be very difficult but is central to a regime-shift.

Practical lessons of the experiments for SNM

The cases bring out some practical lessons for conducting SNM. One important lesson is the importance of committed partners. In the ASTI case partners were selected on the basis of their commitment to the project, which seems a good idea. The cases also show that ambitions should be balanced: they should be neither too high nor too low. Scaling up should be done in a gradual way. Opposition should be anticipated and pre-empted where possible. This was a lesson from the Bologna experiment with automated zone access control systems that experienced mounting political opposition.

The experiments with the use of more sustainable transport technologies and concepts also show that there are advantages in relying on a learning-by-doing approach or probe and learn strategy. This gives flexibility to a project, and offers an opportunity to react to project outcomes and shifts in the environment. The arguably most successful project described in this book, the introduction of organized car-sharing in Switzerland, relied on a phased approach of sequential decision-making and the setting of goals as an evolving activity. The cases also suggest that learning should be made the key aim of experiments rather than quantitative goals (the number of users or vehicles sold). This suggests that projects are best labelled as experiments (or learning experiments). Calling a project an experiment does also have a disadvantage, namely the actors will perceive it as something temporary which may undermine their support for it. This happened in the Praxitèle experiment.

When designing a new experiment, the actors should seek out and utilize previous relevant experience. They should carry out a kind of niche analysis. Furthermore, the expectations of all partners need to be articulated continuously to ensure co-operation of partner activities.

Lastly and perhaps most importantly, experiments should be used to question underlying assumptions at all levels; these include technology options, technology diffusion strategies, and effects upon patterns of mobility. Most of the experiments failed to do this, which meant that the potential for transformation was insufficiently utilized.

These are some of the lessons that followed from the 13 experiments studied in the SNM project. In addition to the case-studies in the present book, these experiments included the introduction of automated zone access control in Bologna, Italy; the road telematics project MOVE-INFO-REGIO in Germany; 'Rolling Highways' projects in Sweden; and the development of autonomous railroad trucks in Germany. They are described in the SNM workbook that resulted from the European Commission supported research project on 'Strategic Niche Management as a Tool for Transition towards a Sustainable Transportation System'.[11] This workbook deals with a range of key dilemmas to be considered by those involved in experiments with new transport technologies, stories which exemplify these dilemmas, and a workshop programme for applying the SNM perspective in a trial or real situation. The lessons do not contradict or invalidate our model of SNM, but rather provide further substance to it.

SNM as a tool for transition

Transitions take a long time. They are the outcome of a myriad of decisions over an extended period in a changing landscape. They are not a linear process but involve processes of co-evolution that give rise to new 'configurations that work',[12] combining old and new elements in novel ways. Technological transitions are associated with structural change at different levels – of companies, production chains, users and government policies – and are connected with new ideas, beliefs and sometimes even new norms and values. Many of the elements involved in transitions cannot be managed. This raises the question: can regime-shifts be managed? The simple answer to this question is 'no, not in a simple way', all one can do and hope for is to exercise some influence, or leverage, to *modulate* ongoing dynamics.

One way to influence processes is through centralized planning. Historically there has been a lot of planning in transport. Infrastructures were built based on planning decisions. Planning has an important role to play in making the current transport regime more sustainable, by reducing the need for transport, providing for transfer places and special infrastructures for cycling and collective transport means to substitute for individual modes of travel. However, there are limitations to a planning approach. When end states and best means to meet these are not clear, as in the case of sustainable travel, one cannot really use a planning and implementation approach. What one can do instead is to try to bend the development process by judiciously applying economic and/or social incentives and disincentives, so as to make some possible paths more, and others less interesting and feasible. This is what policy-makers have tried to do through, for example, the use of gasoline taxes.

A third approach, besides planning and changing the frame conditions, is to '*float with the co-evolution processes and modulate them*'.[13] This is very close to the previous one, but is more directly oriented to dynamics, to learning and adaptation and the creation of visions and plans to guide private and public decision-making. Here policy-makers engage in a kind of process management, exercising some leverage to socially beneficial developments and putting constraints on less desirable developments in order to bend these in more advantageous directions. The support given to electric vehicles and standards on new vehicle automobile emissions are examples of a modulation attempt, aimed at exploring a new path and bending an existing one. SNM is also an example of a modulation approach and in our view an important one. But can it really do the job?

The cases provide little insight for evaluating SNM as a tool for transition. It is unclear whether the experiments will be instrumental in a regime-shift: it is too early to tell. An alternative way to approach the problem is to ask the reverse question: can there be a regime-shift without SNM? The answer is 'yes', if one defines SNM as a deliberate action to achieve a regime-shift. However, if SNM is seen as the introduction of new technology in society in a probe and learn kind of manner, benefiting from special circumstances offered by the local context, the answer is 'no'. Historical studies of technology development offer support for this belief, as discussed in Chapter 2. In our opinion niches are a necessary component of a regime-shift. They help to create a pathway to a new regime without which there will not be a new regime. Niches act as stepping stones. Of course not all niches will be instrumental in this respect. The niche technologies must have ample room for improvement that allows for cost efficiencies and for branching out. They also must have a synergetic relationship with other developments in technology and markets in order to find new users and capture new domains of applications. This is a second necessary condition. A third condition is that the gap between existing domains of application and new ones should not be too big. And a fourth condition is that the rate of progress of the emerging technology system offering particular services should be greater than that of existing technologies with which it must compete. We thus have four conditions for a regime-shift to occur.

Obviously it is difficult to tell beforehand whether these conditions pertain, but they provide a basis for decision-making about technology systems that are eligible for support through niche management. Of course, the choices may be wrong in the sense that no new path is created and the project fails to bear fruit. The attractiveness of SNM is that one finds this out in a bottom-up, non-distortionary manner by carefully choosing a domain of application for which the technology is already attractive. The costs of discomfort are thus minimized (or carried by a local actor with a special interest) while useful lessons may still be learned. Here SNM as a probe-and-learn strategy differs from strategic planning or control policies based on the achievement of set goals in the sense that it is more reflexive and open-ended. It is aimed at the exploration and creation of new paths by building on developments at the local and supra-local level. SNM is thus *not about pushing possible winners but about testing and identifying prospective winners*. The pushing is done after a period of testing, and there is also an element of control, of limiting side effects. SNM thus combines elements of push and control.

The advantage of SNM is that it is targeted to specific problems and needs connected with the use of new technologies and practices. User experiences are used to inform private investment and government support policies. By carefully choosing an appropriate domain, the costs may be kept low. Windows of opportunity are exploited at the local level while at the same time a transition path may be created to a new and more sustainable system in a non-disruptive way. SNM will help actors to negotiate and explore various interpretations of the usefulness of specific technological options and the conditions of their application.[14] The outcomes of the experiments may be used to fine-tune government support policies and to change the frame conditions. This was done in Mendrisio where the large-scale test informed the expansion of the project to the whole of the canton, and in the case of organized car-sharing in which the Swiss authorities intervened to unite the two car-sharing organizations. In the other cases, government intervention to stimulate the diffusion was limited. This very strongly suggests that technology experiments should be supplemented by niche management policies aimed at stimulating the diffusion and further development of niche technologies.

It should be noted that SNM is not a substitute for existing policies for sustainability, but a useful addition. One cannot do without policies that make sustainability benefits part of economic decision-making. Sustainability is a weak driver for change and path creation, far weaker than economic gain is. The two things have to be reconciled: there should be an economic gain in activities that produce sustainability benefits. Subsidies and other types of positive rewards (such as prices) are a possible route for achieving this; taxes, standards and other penalties are another route. Infrastructure provision is a third route. All routes have a role to play, depending on the circumstances.[15] It should be noted that SNM is not so much an instrument to improve the effectiveness of such policies but a way to *improve the functioning of the variation selection process* by increasing the variety of technology options upon which the selection process operates. SNM contributes to the goal of ecological restructuring by exploring options that go beyond the control of particular pollutants and the adoption of eco-efficiency solutions. It is an example of an 'evolutionary' policy, aimed at deliberately shaping paths, creating virtuous circles of positive feedback through carefully targeted policy interventions, rather than at correcting perceived market failures. It thus helps to overcome the weakness of current environmental policies that have been found to have a marginal influence on innovation.[16] It is not a panacea and does not guarantee success, but this holds true for all instruments.

SNM as a modern tool of governance

Above we positioned Strategic Niche Management as part of a third model of governance, which we called modulation policies. Modulation policies are forward-looking and try to utilize the winds of change and seek to exploit windows of opportunity. Such policies are especially suited when end goals are not clear (because manifold) and when there is uncertainty about the best ways to reach them. There are many types of modulation policies. SNM is just one possible policy but in our view an important one, as it helps to deal with uncertainty about the desirability and costs of new technologies and with opposition from vested interests that often stand in the way of doing something new. SNM may actually enrol companies vested in the *status quo* in the process of niche development. However, these companies (the dinosaurs or elephants) should not be allowed to control the process, given their interest in the *status quo*. For radical change one needs outsiders and entrepreneurs.

SNM is not something completely new. It has been attempted, *avant le mot*, by companies for radical innovations such as optical fibres, cellular telephones, aspartame, and computer axial tomography (CT) scanners.[17] But although some attempts such as the Californian Zero-Emission mandate could be labelled as *de facto* SNM policies, it is a new approach for policy-makers. In our view there is a need for policy-makers to go beyond demonstration projects and to promote user experiments with new technologies.

Different people and organizations may thus be interested in technology experiments and SNM for various reasons: to seize a business opportunity, to alleviate a local problem of unsustainability, or simply to learn. Table 6.2 gives an overview of different actors' motivations to engage in technology experimentation.

The table shows that technology experiments allow for mutual benefits that help various parties to find a common ground to be involved in experiments. On the other hand, it shows that SNM involves difficult tradeoffs.

SNM is not something simple; it involves difficult decisions about the use of protection pressures (avoiding overprotection, finding a balance between protection and selection pressures) and the choice of partnership (strong actors versus outsiders). Suggestions for doing SNM are offered in this book and in Weber and Dorda (1999)[18] but it may be clear that SNM can never be carried out in an instrumental way only. This would undermine the reflexive element. It is a perspective that creates a specific kind of communication processes, with specific contents. It helps to align better the technical and social, and this is clearly missing in the cases analysed. Therefore we cannot

Table 6.2 Actors' reasons for engaging in or supporting technology experimentation

Type of actor	Reasons for engaging in or supporting technology experimentation
Companies	• Learn about the current state of a technology either for supply or use and inform company policies • Be prepared for a shift in market conditions creating a demand for a new technology • Influence public policy by offering a solution to an environmental, economic or other type of problem
Local authorities	• Learn about a new technology and about socio-technical arrangements that may solve a local problem (pollution, nuisance, employment, congestion . . .)
State authorities	• Have society learn about new technology options and facilitate transition processes • Create business • Inform public policies to achieve socially desirable outcomes
Consumers and citizen groups	• Learn about their own consumption patterns and needs • Demonstrate to others sustainable life styles • Contribute to a reduction of environmental impacts
NGOs	• Demonstrate feasibility of sustainable lifestyles in order to get support for other policies • Experiments are vehicles for campaigns

be optimistic about organizing regime-shifts. But we also must not be pessimistic. We can hold a position of moderate optimism at best.

An agenda for SNM research and niche management

Further work remains to be done with regard to SNM. More studies need to be conducted of the role of niches in technological regime-shifts; historical studies may be used to this end. A partially unresolved issue is how to organize protection, by what means, and how the phase-out of protection can be done in an ordered, non-disruptive way. In general one should utilize natural forms of protection offered by a local context but this may not be enough. One also needs sponsors and accompanying measures that change the overall frame conditions for economic decision-making. This became most clear in the Bikeabout experiment.

More research is also needed on the relationship between SNM and state policies, and the relationship between SNM and planning. In general, SNM may be used to inform planning (both transport planning and town planning) while planning may be used to foster niche development processes. These issues are unlikely to be resolved on the basis of careful thinking alone. Practical experience is needed to find answers and to generate examples for decision-makers to study and relate to. This is why we propose that decision-makers engage in niche management as part of a more comprehensive strategy for long-term change.[19] So far technology experiments have been largely oriented towards technology testing. They have contributed little to social learning and to processes of co-evolution. We are arguing, therefore, for experiments that are linked to visions and oriented towards social learning, in which users and other actors are encouraged to rethink their perceived needs and basic assumptions. This is already occurring. Some car manufacturers are currently reconsidering their business, we have already discussed this when we talked about Ford. Public transport providers also seem to be changing their mental models, they are becoming more client-oriented and willing to provide door-to-door services. The vision of intermodal travel is widely shared but so far only a few steps have been taken by societal actors to make this a reality. There is a need for further articulating this vision and acting upon this, which requires investment. SNM may help here, but the link between long-term vision (images) and using SNM for further articulating the vision and exploiting it for short-term policies – is only weakly developed in this book and requires more work. Here one can think of the development of socio-technical scenarios based on the multilevel model of evolutionary change, expounded in Chapter 2.[20] Such scenarios are best made in an interactive way, involving different stakeholders, to benefit from different types of knowledge and to make sure that the findings are utilized by decision-makers.

When doing SNM, it is important that experiments are undertaken not as isolated events but that they are linked to long-term strategies for structural change, for instance as part of transition agendas.[21] This guarantees a better utilization of the experimental results and makes introduction strategy more coherent. SNM should also be expanded to include diffusion policies and policies for exploring structural change through system innovation.

With this book we have brought the approach of Strategic Niche Management to the attention of others: practitioners in business and government, non-governmental organizations, and innovation researchers. We have tried to provide a balanced discussion of SNM. We have not tried to sell it as the unequalled tool for introducing technology in society. Sustainability requires a great deal more than technology experimentation, as the cases in this book demonstrate. The introduction of new technology is just a start, just as this book is a start in thinking about how we may move to sustainable transport and to sustainable practices. We invite readers of the book to join in: to try our approach, to talk about it, to refine and improve it.

Notes

1   For this see Levinthal, D.A. (1998) The slow pace of rapid technological change: gradualism and punctuation in technological change. *Industrial and Corporate Change*, 7(2), pp. 217–247; Schot, J. (1998) The usefulness of evolutionary models for explaining innovation. The case of the Netherlands in the nineteenth century. *History and Technology*, **14**, pp. 173–200; and Hoogma 2000, *op.cit.*

2 See Rip, A. and Kemp, R. (1998) Towards a theory of sociotechnical change, in Rayner, S. and Majone, E.L. (eds.) *Human Choice and Climate Change.* Columbus,Ohio: Batelle Press, pp. 327–399; Poel, I. Van de (1998) Changing Technologies. A comparative study of eight processes of transformation of technological regimes. Ph.D. Thesis. Enschede: Twente University Press, and Schot, J., Lintsen, H. and Rip, A. (eds.) (1998) *Techniek in Nederland in de Twintigste Eeuw,* I, chapter 2, pp. 37–39.

3 Knie, A. *et al.* (1997) *Consumer User Patterns of Electric Vehicles.* Research funded in part by the European Commission, JOULE III. Berlin: WZB.

4 When we talk about mistakes, we do not refer to the outcome of the experiment but to weaknesses in the set-up and running of the experiments.

5 Latour, Bruno (1987) *Science in Action. How to Follow Scientists and Engineers through Society.* Cambridge, Mass: Harvard University Press

6 A description of possible societal embedding strategies to be pursued by companies introducing new products in the market is given in Deuten, J.J., Rip, A. and Jelsma, J. (1997) Societal embedment and product creation management. *Technology Analysis & Strategic Management,* 9(2), pp. 219–236. The strategies revolve around anticipation, the mapping of environments and developments, and dialogue.

7 Rip and Kemp, *op. cit.*

8 This does not hold true for experimentation with technology that poses some kind of hazard, as for instance modern biotechnology does. In such circumstances experiments are often protested against by people who fear that experimental use will be followed by widespread application, well before the dangers are sufficiently known and addressed. In the past, the introduction of new technology often sparked protest from workers, nowadays the situation is different. A UK study on work place industrial relations established that workers affected by technical change generally support its introduction, often strongly so, even to the extent that the introduction of new technology can act as a lubricant for less popular change in organization and working practices. See Daniel, W.W. and Millward, N. (1993). Findings of the workplace industrial relations surveys, in Clark, J. (ed.) *Human Resource Management and Technical Change.* London: Sage, pp. 74–75.

9 Burt, Tim (2000) Ford says the future is green.*Wall Street Journal Europe,* 6 October.

10 For a discussion of transitions and transition management, see Rotmans, Jan, Kemp, René and Asselt, Marjolein van (2001) More evolution than revolution. Transition management in public policy. *Foresight,* 3(1), pp. 15–31. This article is based on the Dutch report by Rotmans, Jan, Kemp, René, Asselt, Marjolein van, Geels, Frank, Verbong, Geert and Molendijk, Kirsten (2000) Transities & Transitiemanagement. De casus van een emissiearme energievoorziening. Eindrapport voor studie 'Transities en Transitiemanagement' van ICIS en MERIT t.b.v. van NMP-4 ('Transition and Transition Management'. Study for the Fourth Dutch National Environmental Policy Plan). October.

11 Weber, M., Hoogma, R., Lane, B. and Schot, J. (1999) *Experimenting with Sustainable Transport Innovations. A Workbook for Strategic Niche Management.* Seville/Enschede: University of Twente

12 Rip and Kemp, *op. cit.*

13 Kemp, R., Rip, A. and Schot, J. (1998) Constructing transition paths through the management of niches, in Raghu, Garud and Karnoe, Peter (eds.) *Path Creation and Dependence.* Mahwah, NJ and London: Lawrence Erlbaum. The approaches are not mutually exclusive. Planning and policies that change the frame conditions will be *part* of the third approach, which is more inclusive. The distinction is not so much based on the instrument choice but on the management or governance philosophy.

14 Weber, Hoogma, Lane, and Schot, *op. cit.*

15 A discussion of the pros and cons of environmental policies and the circumstances in which they are best applied is offered in Kemp, R. (2000) Technology and environmental policy: innovation effects of past policies and suggestions for improvement, in *Innovation and the Environment.* Paris: OECD, chapter 3.

16 A discussion of the pros and cons of different environmental policy instruments, especially

the choice between the use of economic incentive and standards, is offered in Kemp, R. (1997) *Environmental Policy and Technical Change. A Comparison of the Technological Impact of Policy Instruments.* Cheltenham: Edward Elgar and Kemp (2000), *op. cit.*

17 Lynn, Gary S., Morone, Joseph G. and Paulson, Albert S. (1996) Marketing and discontinuous innovation: the probe and learn process. *California Management Journal*, 38(3).

18 Weber, Matthais and Dorda, Andreas (1999) Strategic Niche Management: A Tool for the Market Introduction of New Transport Concepts and Technologies. *IPTS Report*, February, pp. 20– 27.

19 An example of such a policy is transition management, adopted by Dutch environmental policy-makers as a model for working towards a transition in energy, agriculture and transport, See Rotmans, Jan, Kemp, Réne and Asselt, Marjolein van (2001) More evolution than revolution. Transition management in public policy. *Foresight*, 3(1), pp 15–31, and Kemp, Réne and Rotmans, Jan (2001) The management of the Co-Evolution of Technical, Environmental and Social Systems. Paper for the International Conference 'Towards Environmental Innovation Systems'. September, Garmisch Partenkirchen, Germany.

20 In socio-technical scenarios paths of co-evolution are explored, using a multi-level perspective of niches, regimes and the socio-technical landscape. Work of this nature is currently undertaken by Frank Geels, Peter Hofman and Boelie Elzen in two Dutch research projects: PRET (Environmental Policy, Economic Restructuring and Endogenous Technology: A Dynamic Policy Analysis) and MATRIC (Management of Technology Responses to the Climate Change Challenge).

21 The concept of transition agendas is described in Rotmans *et al.* (2001) and worked out for the case of a low-emission energy supply system.